Lonsdale STUDENT WORKSHEETS

These student worksheets cover levels 3 - 7 of the revised National Curriculum Study for Key Stage 3. They have been extensively modified, in order to reduce the number of questions which require merely recall, rather than application of facts.

As before, each worksheet is matched specifically to a page in our bestselling Key Stage 3 Revision Guide: 'The Essentials of Science: Key Stage Three' by Katie Whelan.

IDEAS FOR USING THESE WORKSHEETS ...

- As relatively EASY CLASSWORK SHEETS where pupils use their revision guide to provide the answers.

- As HARDER CLASSWORK SHEETS where pupils study the topic first, then answer the questions without their guides.

- As EASY-TO-MARK HOMEWORK SHEETS which provide structured questions similar in style to those in the National Curriculum Tests (many of them can be self-marked by the pupils).

- As TEST MATERIAL to test learning homeworks or indeed to test entire topics.

- As a STRUCTURED REVISION PROGRAMME in the months preceding the National Curriculum Tests.

  and last but not least ...

- As an EMERGENCY MEASURE for when staff are absent at short notice.

- Students should be encouraged to fill in their score at the bottom of each page in the small grey box ☐, and also in the appropriate space in the contents section.

See pages 78, 79 and 80 for information on this guide and our 3 coursebooks for years 7, 8 and 9. Plus information on our websupport worksheets.

© 2000 LONSDALE SRG. ALL RIGHTS RESERVED. NO PART OF THIS PUBLICATION MAY BE REPRODUCED, STORED IN A RETRIEVAL SYSTEM, OR TRANSMITTED IN ANY FORM OR BY ANY MEANS, ELECTRONIC, MECHANICAL, PHOTOCOPYING, RECORDING, OR OTHERWISE WITHOUT THE PRIOR WRITTEN PERMISSION OF LONSDALE SRG.

CONTENTS 1

Page No.

LIFE PROCESSES AND LIVING THINGS

4   Life Processes and Cells
5   Specialised Cells
6   Food
7   A Balanced Diet
8   Digesting Food
9   Enzymes, Absorption and Transport
10 Respiration
11 The Lungs and the Effects of Smoking
12 Bones, Joints and Muscles
13 Adolescence and Reproduction
14 Fertilisation and Development
15 Substance Abuse
16 Microbes and Disease
17 Defence Against Infection
18 Photosynthesis I
19 Photosynthesis II
20 Variation
21 Selective Breeding
22 Classification I
23 Classification II
24 Adaptation to Environment I
25 Adaptation to Environment II
26 Factors Affecting Population Size
27 Food Chains and Food Webs
28 Pyramids of Numbers

MATERIALS AND THEIR PROPERTIES

29 Particle Theory
30 Properties of Solids, Liquids and Gases
31 Change of State
32 Diffusion and Gas Pressure
33 Elements
34 Compounds
35 Mixtures
36 Solutions
37 Separating Mixtures I - Filtration and Evaporation
38 Separating Mixtures II - Distillation and Chromatography
39 Rocks

CONTENTS

MATERIALS AND THEIR PROPERTIES

Page No.

40	Recycling Rocks
41	Recycling Rocks
42	Chemical Reactions
43	Burning Fossil Fuels
44	Reactions of Metals
45	The Reactivity Series
46	Displacement Reactions
47	Acids and Alkalis
48	Reactions of Acids
49	Neutralisation
50	Acid Rain

PHYSICAL PROCESSES

51	Basics of Electricity
52	Electric Current
53	Transferring Electrical Energy
54	Magnetic Fields
55	Electromagnets
56	Balanced and Unbalanced Forces
57	Forces in Action
58	Speed
59	Pressure
60	Moments
61	Light
62	Reflection of Light
63	Refraction and Dispersion of Light
64	Colour
65	Hearing Sound
66	Vibration and Sound
67	The Solar System
68	The Earth
69	Satellites
70	Forms of Energy
71	Energy Resources
72	Temperature and Heat
73	Conduction of Heat
74	Convection, Radiation and Evaporation of Heat
75	Transfer and Conservation of Energy

LIFE PROCESSES AND CELLS

Life Processes and Living Things 1

1. The seven life processes are common to some extent to all life forms.

    All seven are listed below, but the letters have all been mixed up.
    Unjumble the words and then write an explanation for each one.
    1. EEOMVMNT   2. OIRSERPIATN   3. WGTHRO   4. NUDERPROCTOI
    5. IOTXCREEN   6. NOITITRNU   7. TINIESSTIVY

LIFE PROCESS	EXPLANATION
1	
2	
3	
4	
5	
6	
7	

2. Discuss the differences between a pet rabbit and a 'cyber' pet. Why is one living and the other not? Refer to the seven life processes in your answer.

3. The diagram shows a plant cell. Some parts are named.

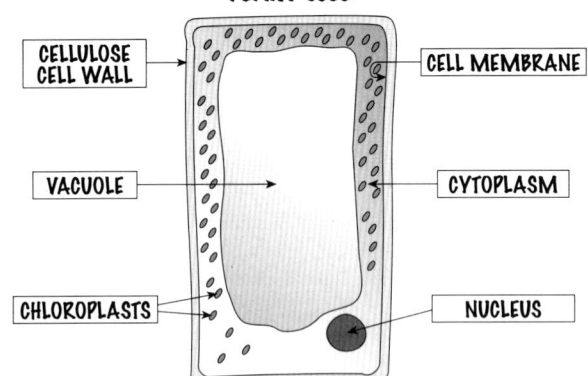

PLANT CELL

CELLULOSE CELL WALL — CELL MEMBRANE — VACUOLE — CYTOPLASM — CHLOROPLASTS — NUCLEUS

a) Which 3 named parts are present in both plant and animal cells?

i) .... ii) .... iii) ....

b) Which 2 named parts are present in plant cells but not animal cells?

i) .... ii) ....

4. For the parts of cells shown, draw a line to match them to their function (job). One has been done.

CELLULOSE CELL WALL	CONSISTS OF CELL SAP UNDER PRESSURE
VACUOLE	PHOTOSYNTHESIS TAKES PLACE HERE
CHLOROPLASTS	CONTROLS WHAT THE CELL DOES AND WHAT IT IS
CELL MEMBRANE	HERE SUBSTANCES ARE MADE AND/OR BROKEN DOWN
CYTOPLASM	PROVIDES SUPPORT FOR THE CELL
NUCLEUS	CONTROLS MOVEMENT INTO AND OUT OF THE CELL

SPECIALISED CELLS

Life Processes and Living Things — 2

1. The list below consists of cells, tissues and organs. Complete the table by putting each word from the list in the correct column.

Brain   Muscle   Sperm   Muscle Tissue   Root Hair Cell   Liver
Blood   Heart   Neurone   Nervous Tissue   Ovum

CELLS	TISSUES	ORGANS

2. The drawings show some plant and animal cells. Each has a different function.
   a) (i) Write down the name of each cell.   (ii) Draw a line from the cell name to its function. One has been done for you.

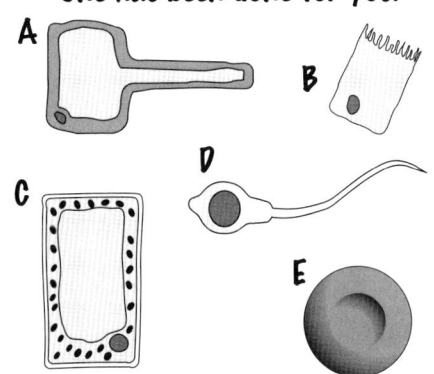

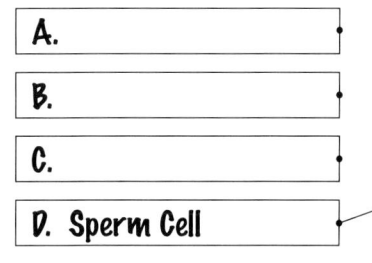

CELL NAME
A.
B.
C.
D. Sperm Cell
E.

CELL FUNCTION
Photosynthesis
Movement of Mucus
Reproduction
Absorption of Water
Carrying Oxygen

   b) (i) What is the function of the head of the sperm cell?

   (ii) What is the function of the tail of the sperm cell?

3. a) Skin cells are very flat and thin and join together at the edges with other skin cells. Explain how skin cells are 'well adapted to their function.'

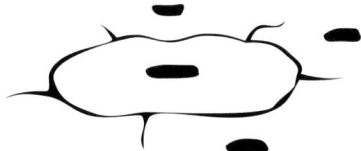

   b) Cells which carry water from the roots of a plant to all the other regions are called XYLEM tissue, and are long and thin. Explain how this type of cell is well adapted to its function.

Key Stage 3 Reference: Page 5   *Lonsdale* Science Revision Guides – The Essential Series   5

FOOD

Life Processes and Living Things — 3

1. Look at the diagrams of foods shown opposite.

   a) Which of these is the best source of vitamin C?

   b) Which of these is the best source of protein?

   c) Which of these is the best source of calcium?

   d) Which of these consists mainly of fat?

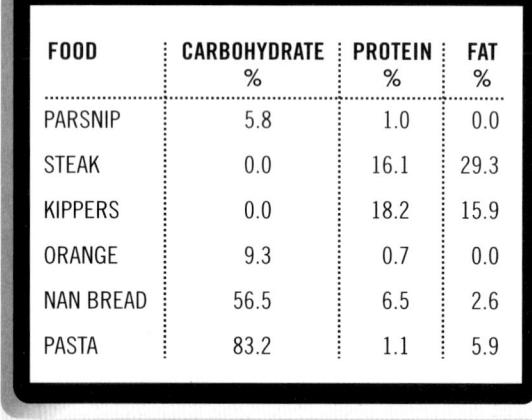

2. Carbohydrates, proteins and fats are three of the nutrients our bodies need.

   a) Name three other nutrients that your body needs.

   (i)   (ii)   (iii)

   Look at the information on the computer screen alongside.

   b) Which food would be most useful to the body for growth and repair?

   c) Which food would contain the least energy?

   d) Which two foods would provide the most energy?

   (i)

   (ii)

FOOD	CARBOHYDRATE %	PROTEIN %	FAT %
PARSNIP	5.8	1.0	0.0
STEAK	0.0	16.1	29.3
KIPPERS	0.0	18.2	15.9
ORANGE	9.3	0.7	0.0
NAN BREAD	56.5	6.5	2.6
PASTA	83.2	1.1	5.9

   e) The figures for each food don't add up to 100%. Which two nutrients would make up for most of the difference?

   (i)   (ii)

3. Look at the information in the bar charts opposite showing the relative amounts of vegetable products and animal products eaten in India and Britain. Explain why cancer of the large intestine is more common in Britain than in India.

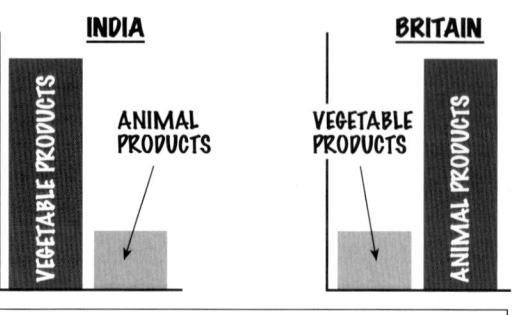

A BALANCED DIET

Life Processes and Living Things — 4

1. a) Explain what is meant by a 'balanced diet'.

   b) Ethan claims that steak in a bun and chips is a balanced diet. The nutritional content of steak, bread and oven chips is shown.

NUTRITIONAL CONTENT PER 100g (g)	Chips	Steak	Bread
Protein	6	27	8
Carbohydrate	24	1	53
Fat	8	23	2
Fibre	1	negligible	2
Water			

   (i) For each type of food work out how much water there is in 100g. Put your answers in the table.

   (ii) Which important nutrients are not listed?

   (iii) If Ethan eats 100g of chips and 100g of steak and 100g of bread for every meal, why is this not a balanced diet?

   (iv) Why is it important that Ethan eats fresh fruit and vegetables?

   c) Ethan is a farmer. Ethan's friend George is a computer operator. Why might they need different diets?

   d) Ethan's wife, Milly, is pregnant. What diet would you recommend to her?

   e) Ethan's baby will need a lot of protein and calcium when it is born.
   (i) Explain why the baby needs protein and calcium.

   (ii) Name a good food source of protein and calcium.

   f) Ethan decided to run a marathon. Before he ran the marathon he ate a lot of pasta.
   (i) What is the main constituent of food in pasta?

   (ii) Why was it important for Ethan to 'load up' with pasta?

Key Stage 3 Reference: Page 7 — *Lonsdale* Science Revision Guides - The Essential Series

DIGESTING FOOD

Life Processes and Living Things — 5

1. Fill in the gaps in the following passage ...

   Food is mainly made up of .... molecules which are often ....
   These molecules must first be .... by .... in the digestive system so that
   they become .... and .... This makes it easy for them to pass through the
   walls of the .... and into the .... The three main food types which need
   to be .... in this way are ...., ... and ....

2. a) Why is it necessary for food to be digested?

   b) "A molecule like glucose can be absorbed immediately by the body, but a molecule like starch takes much longer." Use your knowledge of digestion to explain a possible difference between the two molecules.

   c) Some molecules manage to pass through the digestive system without being digested. Use your knowledge of digestion to describe one possible feature of this type of molecule.

3. Match the start of these sentences about the digestive system with their correct endings.

The pancreas ...	... produces an enzyme which acts on protein.
The small intestine ...	... contains an enzyme which acts on starch.
The large intestine ...	... pours enzymes into the small intestine.
The mouth ...	... reabsorbs excess water from the food mixture.
The stomach ...	... absorbs the products of digestion.
Enzymes ...	... increase the surface area of food by breaking it up.
The teeth ...	... are proteins which break down food.

ENZYMES, ABSORPTION AND TRANSPORT

Life Processes and Living Things — 6

1. Imran and Jack carried out an investigation to see how the digestion of starch is affected by an enzyme. The results are shown in the table. If starch was present the iodine went blue. If starch was not present the iodine stayed brown.

		\multicolumn{7}{c}{Time (min)}
		0	5	10	15	20	25	30
Temperature (°C)	20	Bl	Bl	Bl	Bl	Br	Br	Br
	35	Bl	Bl	Br	Br	Br	Br	Br

Bl = Blue   Br = Brown

a) Why did the mixture stop turning the iodine blue?

b) At what temperature did the enzyme work best?

c) Why do you think it works best at this temperature?

d) The diagram shows an experiment to show how the human gut works. Starch solution is placed inside the visking tubing. After 20 minutes the water surrounding the visking tubing is tested for starch.

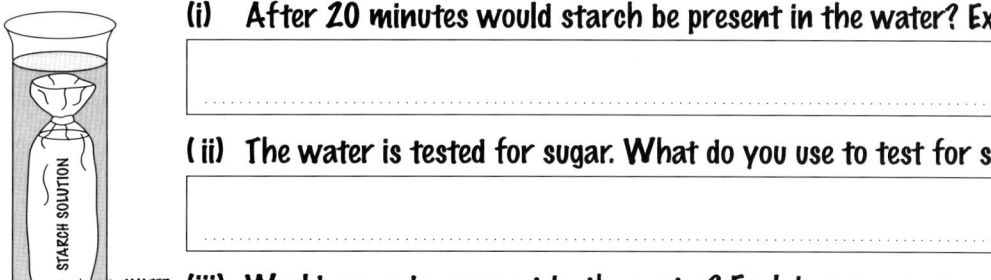

   (i) After 20 minutes would starch be present in the water? Explain your answer.

   (ii) The water is tested for sugar. What do you use to test for sugar?

   (iii) Would sugar be present in the water? Explain your answer.

e) The experiment is repeated but this time amylase is added to the starch in the tubing.
   (i) The water is again tested after 20 minutes. Would starch be present?

   (ii) The water is tested for sugar. Would sugar be present? Explain your answer.

f) Describe how the products of digestion get from inside the small intestine to the cells in the rest of the body.

RESPIRATION

Life Processes and Living Things — 7

1. Sunil is out walking. He gets his energy from the process of respiration.

   a) What do we mean by 'aerobic respiration'?

   b) Write down a word equation for aerobic respiration.

   c) The statements below describe simply some of the processes involved in respiration. Put them into the correct order by putting the letters into the boxes below.

   A Oxygen from the lungs diffuses into the blood.

   B Carbon Dioxide diffuses back into the air sacs in the lungs.

   C Oxygen diffuses into the muscle cells.

   D Oxygen is used in respiration to produce energy, carbon dioxide and water from glucose.

   E Carbon Dioxide is transported back to the lungs

   F Oxygen is transported in the blood to all parts of the body.

   G Carbon Dioxide from respiration diffuses back into the blood.

   Oxygen is breathed into the lungs → ☐ → ☐ → ☐ → ☐ → ☐ → ☐ → ☐ → Carbon Dioxide is breathed out

   d) Explain how the glucose, used in respiration, is transported to the muscle cells.

   e) Sunil decides to run home. He begins to breathe deeper and his pulse rate increases. Explain why this is so.

   f) Sunil feels thirsty when he gets home. State 3 ways in which he loses water from his body.

   (i)   (ii)   (iii)

   g) Explain the difference between respiration and breathing.

THE LUNGS AND THE EFFECTS OF SMOKING

Life Processes and Living Things — 8

1. Look at the diagram of an air sac and answer the following questions.

   a) What difference would you expect to find in the blood at points A and B?

   b) Which gas is shown leaving the blood at point C?

   c) Which gas is shown entering the blood at point D?

   d) The diagram below shows a damaged air sac as a result of emphysema.

   (i) The amount of gas exchanged in a damaged air sac is less. Explain why.

   (ii) Emphysema is caused by smoking. Explain why.

   e) Different substances in cigarette smoke have different effects.

   (i) Which part causes addiction to smoking cigarettes?

   (ii) Which part may cause lung cancer?

   (iii) Which part reduces the oxygen carried by red blood cells?

   (iv) Which part causes bronchitis?

   f) Explain why smokers often get out of breath very easily.

2. Atmospheric air contains 21% oxygen and 0.03% carbon dioxide.

   a) How much oxygen and carbon dioxide is there in the air that we breathe out?

   b) How do you explain the difference between air breathed in and air breathed out?

BONES, JOINTS AND MUSCLES

Life Processes and Living Things — 9

1. The diagram opposite shows a typical joint.

   a) Label A and B.

   A ....   B ....

   b) What is the job of ...

   (i) the cartilage? ....

   (ii) the part labelled A? ....

   (iii) the part labelled B? ....

2. a) Describe a way in which our skeleton could improve our survival chances if we were involved in a road accident.

   b) Describe a way in which our skeleton helps us to throw a ball.

3. a) What type of joint is shown in these diagrams?

   b) Why is it important that some joints in the body have a very limited range of movement?

4. The diagram shows part of the arm.

   a) What is the name given to the part of the body labelled C that attaches muscles to bones?

   b) Describe how each of the following movements can be brought about.
   (i) Bending the arm.

   (ii) Straightening the arm.

   c) What is the name given to two muscles acting in this way?

12 — Lonsdale Science Revision Guides - The Essential Series — Key Stage 3 Reference: Page 12

ADOLESCENCE AND REPRODUCTION

Life Processes and Living Things — 10

1. Boys and girls go through changes during adolescence.

   a) List the changes which occur in girls.

   b) Name a change which occurs in both girls and boys.

   c) Why do these changes happen?

2. Label the diagram of the female reproductive system. Briefly describe the function of each part.

   A ....
   B ....
   C ....
   D ....
   E ....

3. Label the diagram of the male reproductive system. Briefly describe the function of each part.

   A ....
   B ....
   C ....
   D ....

4. Complete the stages by which an egg (ovum) may be fertilised through sexual intercourse.

   (i) Sperm is continually produced in the testes.
   (ii) An egg is released monthly and moves into the oviduct.
   (iii) ....
   (iv) ....
   (v) ....
   (vi) ....
   (vii) ....
   (viii) FERTILISATION may occur if a sperm then fuses with an egg.

Key Stage 3 Reference: Page 13 — Lonsdale Science Revision Guides - The Essential Series

FERTILISATION AND DEVELOPMENT
Life Processes and Living Things — 11

1. Between the ages of approximately 13 and 50, a woman is fertile and the lining of her uterus is replaced every month. Assuming that the lining of the uterus starts to break down on day 0, ...

   a) For how long (approximately) does the lining of the uterus continue to break down?

   b) On what day (approximately) is the egg released from the ovary?

   c) If fertilisation doesn't occur, on what day (approximately) would the lining next start to break down?

   d) For how long (approximately) does the wall of the uterus remain at its thickest?

   e) How long (approximately) does it take for the uterus wall to regain its full thickness after a period?

2. Arrange the following sentences into their correct order.

   A The egg moves into the fallopian tube and starts to move down it.
   B The fertilised egg now starts to divide rapidly to form a ball of cells.
   C If intercourse has taken place, then sperm may have swum up into the fallopian tubes.
   D When the ball of cells reaches the uterus it implants into its spongy lining.
   E If a sperm fuses with the egg, then fertilisation is said to have happened.
   F Once a month an egg is released from an ovary.
   G From here it continues to develop into an embryo.

3. The diagram shows a baby developing inside its mother's uterus.

   a) What difference will there be in the oxygen content of the blood at points A and B?

   b) What differences will there be in the carbon dioxide content of the blood at points A and B?

   c) What other differences would you expect to find in the blood at points A and B?

   d) The baby is surrounded by amniotic fluid in the uterus. What is the purpose of this fluid?

SUBSTANCE ABUSE

Life Processes and Living Things — 12

1. Complete the crossword below ...

   ACROSS
   1. Tar in cigarette smoke causes this terrible disease (4,6).
   6. A very important organ which can be permanently damaged by solvent abuse (5).
   9. Even mild cigarettes contain this substance (3).
   10. Alcohol and barbiturates are examples of these (11).
   11. A major body structure which can be damaged by solvent abuse (5).
   12. 'Speed' and Methedrine are examples of these (10).
   13. 'Seeing things that aren't there'. Can be caused by L.S.D. (13)

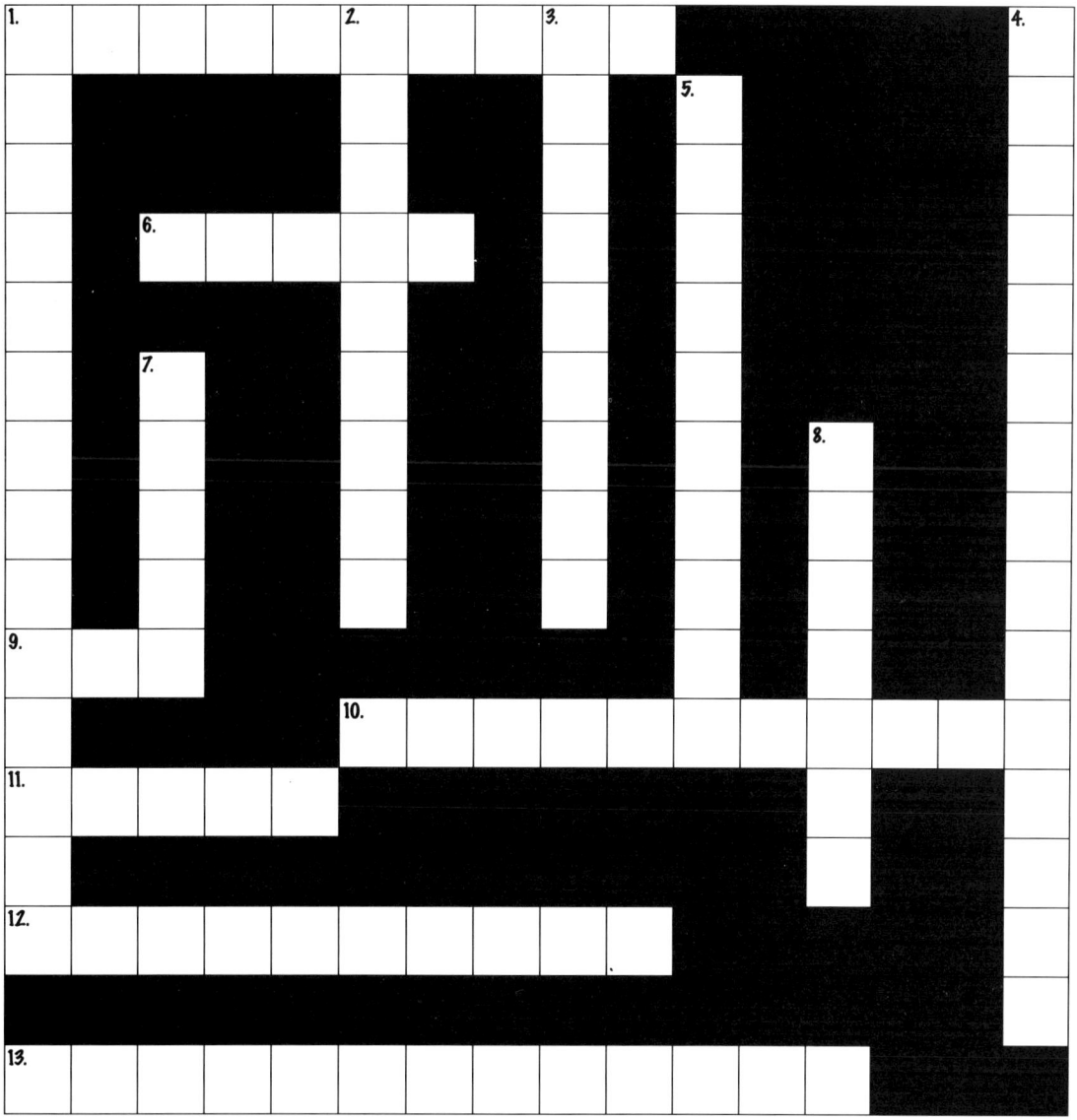

   DOWN
   1. These are more common amongst smokers (4,10).
   2. This occurs when you are absolutely dependent upon something (9).
   3. A lung condition in which the air sacs are damaged through coughing (9).
   4. This condition, caused by smoking can lead to heart attacks and strokes (8,7).
   5. Irritation of the lungs caused by smoke and excess mucus (10).
   7. Largest organ of the body which can be permanently damaged by alcohol abuse (5).
   8. An hallucinogenic drug which can lead to dehydration and collapse (7).

MICROBES AND DISEASE

Life Processes and Living Things — 13

1. Flu and pneumonia are both caused by microbes. Flu is caused by a virus and pneumonia by a bacterium.

   a) Name one difference and one similarity between the flu virus and the pneumonia bacterium.

   BACTERIA

   VIRUSES — genes, protein coat

   b) Why can't you treat flu with antibiotics?

2. Which of the following diseases could you treat with antibiotics? Underline your answers.

   Tetanus   Colds   Flu   Cholera   Measles   Tuberculosis   Polio

   Athlete's Foot   Whooping Cough   Aids   Hayfever   Salmonella

3. a) Why is it important that frozen meat, like chicken, is thawed out and cooked properly?

   b) Athlete's foot fungus likes warm damp conditions. Why is it common to catch athlete's foot in the changing rooms at the baths?

   c) Josie caught a cold. 'But I haven't touched anyone with a cold' said Josie.
   Why isn't it necessary for Josie to have touched anyone?

   d) First aiders wear rubber or plastic gloves. Why is this important for ...

   (i) the first aider

   (ii) the person being treated

   e) Syphilis is a venereal disease. Explain how syphilis is spread.

   f) Before travelling to some parts of Africa people are given advice ...
   'have a malaria jab'   'drink only bottled water'
   (i) How would the person contract malaria?

   (ii) Why are they advised to drink only bottled water?

DEFENCE AGAINST INFECTION

Life Processes and Living Things — 14

1. Phil was playing football, he was tackled badly and cut his leg. His manager asked him if he'd had a tetanus jab. Phil said 'no.'

   a) Tetanus is a type of bacteria, describe how they could enter Phil's body and make him feel ill.

   b) Describe how the white blood cells in Phil's body will try to protect him from the tetanus bacteria.

   c) A 'tetanus jab' could immunise Phil against the tetanus. This is carried out by injecting a dead form of the bacteria into Phil's body. Explain how immunisation will help prevent Phil from getting ill from the tetanus bacteria.

2. Poliomyelitis (polio for short) is a viral disease which attacks the spinal cord. It was widespread throughout Europe and America. In the 1950s an immunisation programme was launched. Now polio is almost unknown in developed countries.

   a) If someone were to develop polio describe how the virus would have reproduced in their body.

   b) Could polio be treated with antibiotics? Explain your answer.

   c) Polio is now almost unheard of in developed countries. Does this mean that the polio virus does not exist any more? What would happen if the immunisation programme ended tomorrow?

3. Some diseases such as measles we get once and never get again. We have acquired a natural immunity to the disease.

   Explain what we mean by 'natural immunity.'

Key Stage 3 Reference: Page 17 — Lonsdale Science Revision Guides - The Essential Series

PHOTOSYNTHESIS I

Life Processes and Living Things — 15

1. The diagram shows an experiment to show the effect of light intensity on the rate of photosynthesis of a green plant, pond weed. The light intensity is altered by moving the lamp towards or away from the pond weed. The rate of photosynthesis was measured by counting the number of bubbles of gas given off every minute.

   a) Name the gas given off in photosynthesis.

   b) Write down the word equation for photosynthesis.

   _____ + _____ —LIGHT→ _____ + _____

   c) The graph opposite shows the results of the experiment.
      (i) As the lamp is moved away from the pond weed does the light intensity decrease or increase?

      (ii) What happens to the number of bubbles per minute as the light intensity increases?

      (iii) What happens to the rate of photosynthesis as the light intensity gets greater?

   d) The graph opposite shows how the amount of oxygen given off by a tree each day, changes over a year.
      (i) Describe how the rate of photosynthesis changes over the year.

      (ii) Give two reasons why photosynthesis occurs most rapidly in summer.

      1.
      2.

2. The diagram shows a variegated leaf. The area marked A is green and area B is white.

   a) What substance is present in area A but not B?

   b) In which area would you expect glucose to be produced in photosynthesis?

3. Describe two ways in which the leaf is adapted for photosynthesis?

   1.
   2.

PHOTOSYNTHESIS II

Life Processes and Living Things — 16

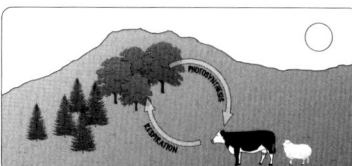

1. All living things respire aerobically in order to release energy from food.

   a) Which substance is used up during aerobic respiration?

   b) Which substance is released as a waste product at the same time?

   c) Which substance is used up during photosynthesis?

   d) Which substance is released as a waste product at the same time?

2. Why is it important for the planet that the processes of respiration and photosynthesis stay in balance?

3. The world's forests are being chopped down at an alarming rate.

   a) What effect will this have on the balance between photosynthesis and respiration?

   b) Besides leaving the trees alone, what else could be done to try to "keep the balance"?

4. Can you explain the shape of the root hair cell drawn below?

5. Why can too much water be as bad as no water for plant roots?

6. Which 3 minerals are essential for good plant growth? State why they are needed.

MINERAL	WHY NEEDED
1	
2	
3	

Key Stage 3 Reference: Page 19

VARIATION

Life Processes and Living Things — 17

1. Some pupils collected data from everybody in the class about ...
   i) their eye colour and ii) their shoe size. The results are shown below.

   a) Plot graphs to show the results.

   (i) Eye colour.

Eye colour	Blue	Green	Brown	Hazel
No. of pupils	10	4	14	2

   (ii) Shoe size

Shoe size	5	6	7	8	9	10
No. of pupils	2	3	6	8	5	2

   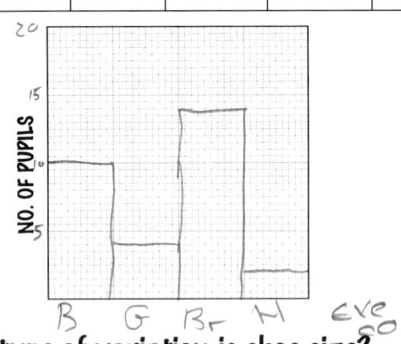
   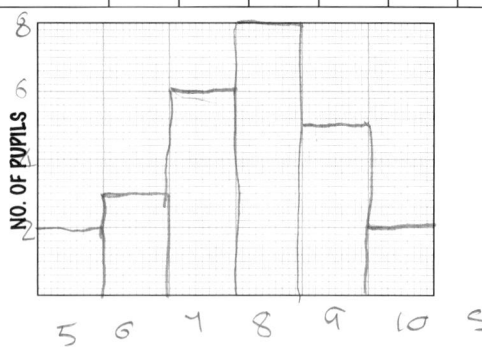

   b) What type of variation is shoe size?

   c) What type of variation is eye colour?

   d) For the following characteristics state whether they are caused by genetic factors, environmental factors or both factors.

Example of variation	Cause	Example of variation	Cause
Sex	genetic	Eye colour	genetic
Height	both	Weight	environment
Speed	both	Dyed blonde hair	environment
Strength	environment	Beard thickness	environment
Cut on face	environment	Attached ear lobes	genetic
Suffer from haemophilia	both	Intelligence	both
Got a bad cold	environment	Blood group	genetic

2. The two graphs show the height of pupils in two year 9 classes. One class comes from France and the other from a famine stricken area of Africa.

   a) Which set of data comes from France?

   b) Suggest reasons why the data is different for each group of students.

   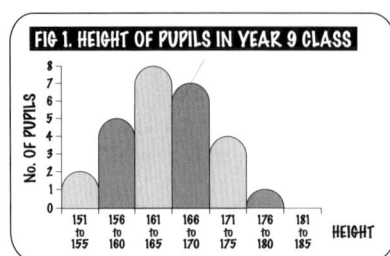
   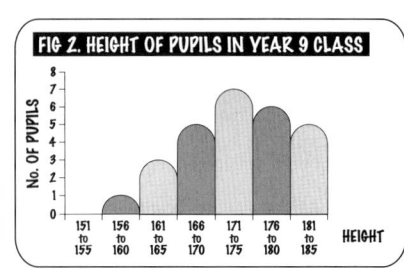

   FIG 1. HEIGHT OF PUPILS IN YEAR 9 CLASS

   FIG 2. HEIGHT OF PUPILS IN YEAR 9 CLASS

SELECTIVE BREEDING

Life Processes and Living Things — 18

1. Explain what is meant by selective breeding.

2. Fran has several Jersey cows. Their yields are shown. Fran plans a breeding programme to get cows that give larger amounts of milk every year.

Name of Cow	Yield (litres per year)
Becky	4500
Michelle	3500
Daisy	4000
Andrea	3000

   a) Which cow should she choose to mate with Brian the Bull at the beginning of the breeding programme? Explain your choice.

   b) How should Fran choose cows from the offspring to continue the programme?

   c) Jersey cows are renowned for their cream but give low yields. Holstein cows give high yields but little cream. Describe how you would breed a cow to give high yields of creamy milk?

   d) Cotswold sheep have long woolly coats. Suffolk sheep give lots of meat. Describe how you could breed a sheep that would produce lots of wool and lots of meat.

3. Most grapes sold these days are of the seedless variety. Thirty years ago seedless grapes were unheard of.

   a) Why have grape growers developed the seedless grape.

   b) Suggest how the seedless grape might have been developed.

   c) Modern vegetables took hundreds of years to develop from their ancestors. Why was it possible to develop the seedless grape in only 30 years or so? (assume no genetic modification [GM]).

4. Golden retriever dogs often have problems with ear infections. Suggest ways in which, if at all, selective breeding could get rid of this characteristic.

Key Stage 3 Reference: Page 21 — *Lonsdale* Science Revision Guides - The Essential Series

CLASSIFICATION I

Life Processes and Living Things

1. The diagram below shows different groups of animals. Fill in the blank boxes to complete the diagram.

```
                          ANIMALS
            ┌────────────────┴────────────────┐
        VERTEBRATES                      INVERTEBRATES
      ┌─────┬──┴──┬──────┐              ┌───────┬────┴────┐
    [   ] REPTILES [   ]              [   ]          MOLLUSCS
      │              │                   ├────┬────┐        │
    BIRDS         [   ]                [   ]    [   ]     [   ]
                                                 │
                                              ARACHNIDS
```

2. The Woolly Mammoth was a large mammal.

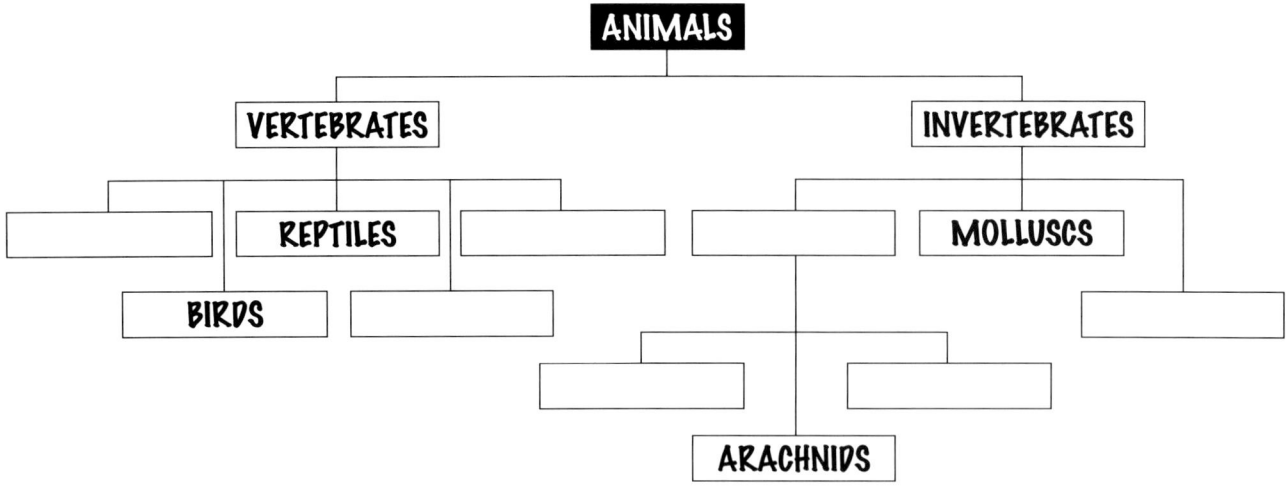

   a) How can you tell from the drawing that the Woolly Mammoth was a mammal?

   b) The Woolly Mammoth is a vertebrate. What does vertebrate mean?

   c) Mammals are different from other vertebrates. State one way in which mammals differ

   d) Dinosaurs are now extinct. They were reptiles. Describe the characteristics of a dinosaur.

3. Sebastian took a notebook and pencil and went walking in the country. He jotted down notes about some of the animals he saw. Some of his descriptions are written down here. Which animal group do they belong to?

   (A) - body divided into segments, bristles
   (B) - body in 3 parts, gills, jointed legs
   (C) - external skeleton body in 3 parts, wings
   (D) - body divided into 2 parts, no wings
   (E) - hard shell, no body segments
   (F) - body in 3 parts, 3 pairs of legs

   A =
   B =
   C =
   D =
   E =
   F =

CLASSIFICATION II

Life Processes and Living Things — 20

1. For the following green plants draw a line to link the group to its description.

   PLANT GROUP:
   - MOSSES
   - FERNS
   - CONIFERS
   - ALGAE
   - FLOWERING PLANTS

   DESCRIPTION:
   - No roots, stems or leaves.
   - Found in water.

   - HAS roots, stems and leaves.
   - Reproduce by SPORES.

   - No ROOTS, simple leaves and stem.
   - Reproduce by SPORES.

   - Produce POLLEN and/or EGG CELLS.
   - Seed develops from part of flower.

   - NEEDLE-LIKE leaves.
   - Produce SEEDS on CONES.

2. Seaweeds are algae, hartstongue is a fern, Scots Pine is a conifer and daffodil is a flowering plant. They are all part of the plant kingdom.

   a) Explain the difference between spores and seeds.

   b) Explain what is meant by a vascular system.

   c) Discuss similarities and differences in the characteristics of:

   (i) Seaweed and Scots Pine

   (ii) Hartstongue and daffodil.

   (iii) Seaweed and daffodil.

   (iv) Hartstongue and Scots Pine.

   (v) Scots Pine and daffodil.

   (vi) Seaweed and hartstongue.

Key Stage 3 Reference: Page 23

ADAPTATION TO ENVIRONMENT I

Life Processes and Living Things — 21

1. Describe the habitat in which you would expect to find the following organisms.

   (i) Snail .... (vi) Whale ....

   (ii) Pike .... (vii) Caterpillar ....

   (iii) Buttercup .... (viii) Fir Tree ....

   (iv) Badger .... (ix) Zebra ....

   (v) Rabbit .... (x) Rat ....

2. The drawings show different animals. Each one has special features that help it to survive in its habitat. For each one describe the special features and say how these help them to survive.

   a) ....

   b) ....

   c) ....

   d) ....

   e) ....

3. The bluebell is a woodland plant. It flowers in early spring and dies back before summer. Explain how this helps it to survive.

4. Ivy is found on trees in woodlands. It climbs up the trunk. Explain how this helps the ivy to survive.

5. Slugs live in cracks in stone walls. How are they adapted to live there?

ADAPTATION TO ENVIRONMENT II

Life Processes and Living Things — 22

1. Explain how the following animals are adapted to survive.

   a) Snail (prey) ....

   b) Hedgehog (prey) ....

   c) Fox (predator) ....

   d) Cheetah (predator) ....

2. Plants and animals adapt to daily changes in their environment.

   a) Some plants close the stomata on their leaves at night. Why do they do this?

   b) (i) The dormouse is nocturnal. Why does it only come out at night?

   (ii) Owls eat dormice. When would you expect owls to hunt?

3. Hawks hunt for small animals such as voles. They hover about 10 metres above them and then dive swiftly on their prey.

   a) What is the advantage of hunting for voles from such a height?

   b) Suggest 3 different features of hawks which enable them to be successful at catching voles from such a height.

   (i) ....

   (ii) ....

   (iii) ....

4. a) Animals and plants adapt in different ways to the onset of winter. Suggest two ways in which animals can adapt to seasonal changes.

   1. ....

   2. ....

   b) Why do Arctic geese 'winter' in Britain?

Key Stage 3 Reference: Page 25 — *Lonsdale* Science Revision Guides - The Essential Series

FACTORS AFFECTING POPULATION SIZE

Life Processes and Living Things — 23

1. a) The dormouse is becoming extinct in many parts of Britain. It lives in ancient woodland. Suggest reasons why the population of dormice is decreasing.

   b) Owls hunt dormice. Suggest what has happened to the owl population in these areas.

   c) Foxes are found in ancient woodland. Suggest why increasing numbers of foxes are now found in our towns and cities scavenging for food.

2. The population of a species varies over a period of time. The food chain below shows the feeding relationship between some organisms living in England.

   GRASS ⇒ RABBIT ⇒ STOAT

   a) Which organism in the food chain is the predator?

   b) In one area of England the rabbit population has been counted every 5 years since 1945. The results are given in the table below.

NO. OF RABBITS	YEAR
500	1945
1500	1950
2900	1955
3000	1960
1800	1965
1000	1970
1800	1975
3300	1980
2700	1985
1300	1990
1000	1995

   i) Plot the results on the graph paper.

   ii) Sketch on the graph how you would expect the population of stoats to have changed over the same period.

   c) Why might the rabbit population have ...

   i) Increased:

   ii) Decreased:

FOOD CHAINS AND FOOD WEBS

Life Processes and Living Things — 24

1. For the following draw a line to link the right word to the right definition. One has been done for you.

HERBIVORE	Release nutrients from dead animals
DECOMPOSER	They only eat plants
OMNIVORE	They only eat animals
PRODUCER	Green plants that make their own food
CARNIVORE	They eat plants and animals

2. The diagram shows a food web for a wood.

   a) What is the name of the producer in this food web?

   b) From the food web write down two of each of the following:

   (i) OMNIVORES: 1. .... 2. ....

   (ii) PRIMARY CONSUMERS: 1. .... 2. ....

   (iii) SECONDARY CONSUMERS: 1. .... 2. ....

   (iv) TOP CARNIVORES: 1. .... 2. ....

   (v) HERBIVORES: 1. .... 2. ....

   c) From the food web above draw: (i) A food chain consisting of the oak tree and 3 animals.

   (ii) A food chain including blue tits and one other animal.

   d) What would be the effect on the food web above if the greenfly were sprayed with insecticide that killed most of them off? Try to think of SIX possible effects.

1.	2.
3.	4.
5.	6.

   e) Which two animals are in competition with each other for dormice?

   f) What would be the effect on the food web if an owl that ate dormice appeared in the wood?

Key Stage 3 Reference: Page 27

PYRAMIDS OF NUMBERS

Life Processes and Living Things — 25

1. For the following food chains draw a pyramid of numbers.

   a) Blades of Grass - Rabbit - Fox

   b) Algae - Shellfish - Lobster

   c) Microscopic Plants - Microscopic Animals - Minnows - Heron

   d) Rose Bush - Aphids - Ladybirds - Blue Tits - Hawk

   e) Oak Tree - Dormice - Owl

   f) Cabbage - Caterpillars - Thrush - Cat

2. The following diagrams show FOUR different pyramids of numbers. Which diagram - A, B, C, or D best matches the following food chains?

   a) Microscopic plants are eaten by water fleas. Water fleas are eaten by minnows which are eaten by pike.

   b) Rose bush are eaten by greenfly, greenfly are eaten by sparrows, sparrows are eaten by a hawk.

   c) Antelopes feed on long grass, ticks suck the blood of antelopes, birds sit on the antelope's back and eat the ticks.

   d) Zebras feed on long grass, lions feed on zebras, fleas suck the blood of lions.

3. For the food chain shown:

   FOX ← RABBIT ← GRASS

   a) Explain why all the energy contained in the grass is not transferred to the fox.

   b) Why is it more expensive to buy "free range" chicken than "battery" chicken?

---

28 — Lonsdale Science Revision Guides - The Essential Series — Key Stage 3 Reference: Page 28

PARTICLE THEORY

Materials and their Properties — 1

1. Lead is a solid. Use the particle theory to explain the following properties of lead (as a solid).

   a) Lead cannot be compressed.

   b) Lead does not flow.

   c) Lead has a high density.

   d) Lead always stays the same volume and shape.

   e) Draw a diagram to show the arrangement of the particles in lead.

2. Oil is a liquid. Use the particle theory to explain the following properties of oil.

   a) Oil cannot be compressed.

   b) Oil can flow.

   c) Oil has a medium density.

   d) Oil stays the same volume but it alters its shape to fit the bottom of the container it is put in.

   e) Draw a diagram to show the arrangement of the particles in oil.

3. Air is a gas. Use the particle theory to explain the following properties of air.

   a) Air can easily be compressed.

   b) Air can flow.

   c) Air has a very low density.

   d) Air takes the shape of the entire container it is put in.

   e) Draw a diagram to show the arrangement of the particles in air.

Key Stage 3 Reference: Page 29 — *Lonsdale* Science Revision Guides - The Essential Series

PROPERTIES OF SOLIDS, LIQUIDS AND GASES

Materials and their Properties — 2

1. Below are TWELVE statements which describe a property of a solid, a liquid or a gas. Link each property to the correct state or states. One has been done for you.

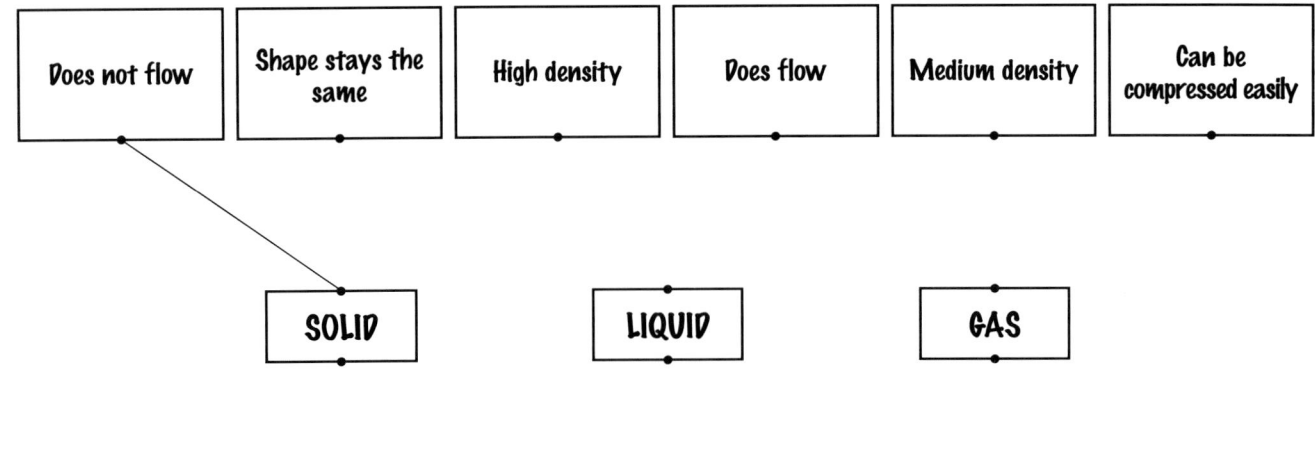

2. What state am I?

   a) I cannot be compressed and I flow?

   b) My volume and shape stays the same?

   c) I can flow and be compressed easily?

   d) My shape does not stay the same and I have a very low density?

3. Explain the following:

   a) Air is put into tyres.

   b) Bike frames are made from metal.

   c) Paints are liquid when applied, but need to be solid on the wall.

   d) Water is distributed in pipes.

   e) A gas leak will quickly fill a room.

CHANGE OF STATE

Materials and their Properties 3

1. For the following changes of state, say whether the substance is:

   MELTING, FREEZING, BOILING or CONDENSING.

   a) SOLID ⟶ LIQUID ....
   b) LIQUID ⟶ GAS ....
   c) GAS ⟶ LIQUID ....
   d) LIQUID ⟶ SOLID ....

2. A beaker of ice was taken out of a freezer and left to melt. Its temperature was taken every minute. The graph shows how the temperature of the ice changed.

   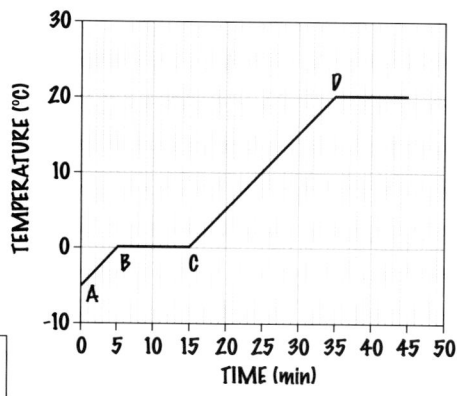

   a) What is happening to the ice between B and C?

   b) i) What state is the ice in between A and B?

   ii) Describe the motion of the particles of ice in this region.

   c) i) What is happening in the region C to D?

   ii) How does the distance between the particles in the region C to D compare to A to B?

   d) Why does the temperature of the water stay constant at point D?

   e) The mass of the beaker of ice was taken at the start and again after 15 minutes. The mass at the start was 300g. What was the mass after 15 minutes? Explain your answer.

3. Martha carried out an experiment where she allowed a liquid to cool. She took its temperature every minute for 10 minutes. The results are shown below.

TIME (min)	0	1	2	3	4	5	6	7	8	9	10
TEMPERATURE (°C)	90	84	79	75	75	75	75	70	65	62	59

   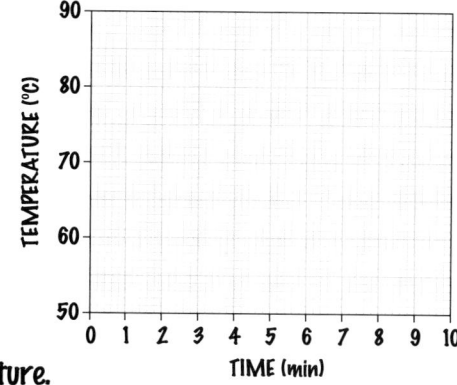

   a) Plot a graph of her results. Draw a line of best fit.

   b) At what temperature does the liquid solidify?

   c) Using the particle theory, describe what happens at this temperature.

Key Stage 3 Reference: Page 31

DIFFUSION AND GAS PRESSURE

Materials and their Properties — 4

1. Explain what is meant by diffusion.

2. Joe has just been doing PE. At the start of his next lesson he sprays himself with deodorant. Marie who is sat at the other end of the classroom sees Joe spray himself but only smells it some time afterwards. Can you explain how the smell reaches Marie? (Joe and Marie do not move).

3. The drawing shows the deodorant can that Joe used. In the can is a liquid and a gas under pressure.

   a) Tick **one** correct statement about the distance between the molecules in a gas, compared to those in a liquid.

   ☐ Closer together than those in a liquid.   ☐ Further apart than those in a liquid.

   ☐ The same distance apart as those in a liquid.

   b) How do moving gas molecules exert a pressure on the inside of the can?

   c) You must never put a pressurised deodorant can near a hot fire. Explain why.

4. a) While Tom is out on a bike ride, he gets a flat tyre. The pressure in the tyre decreases as air leaves the tyre. Explain why.

   b) Tom pumps up the tyre. The air in the tyre is warmed up by the pumping. What affect will this have on ...

   i) The air particles in the tyre:

   ii) The pressure of the air in the tyre:

ELEMENTS

Materials and their Properties — 5

1. The periodic table is a way of classifying elements.
   On the periodic table below, colour all the metals in blue and all the non-metals in red.
   Complete the table by adding the correct symbol for each element name given.

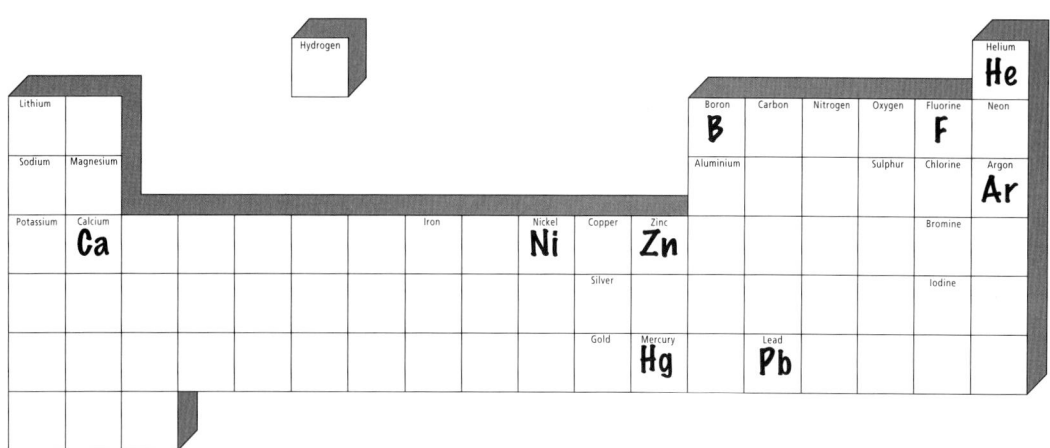

2. The table contains data about the melting points and boiling points of some elements.

	A	B	C	D	E	F	G	H	I	J
MELTING POINT (°C)	660	-218	650	1540	-272	419	-101	961	-7	-249
BOILING POINT (°C)	2470	-183	1110	2750	-269	907	-35	2212	59	-246

a) Write down the letters of the elements which are:

i) Solid at room temperature (20°C): ....

ii) Liquid at room temperature (20°C): ....

iii) Gaseous at room temperature (20°C): ....

b) Write down the letters of THREE metals from the table above.   i) ....   ii) ....   iii) ....

3. From the following descriptions state whether the element is a metal or non-metal.

ELEMENT	DESCRIPTION	Metal/Non-metal?
a	Melting point of -39°C. Boiling point 357°C. Conducts electricity.	
b	Gaseous at room temperature. Does not conduct electricity when liquid.	
c	Solid at room temperature. Magnetic. Conducts electricity as a solid.	
d	Yellow solid at room temperature. Non-magnetic. Does not conduct electricity.	

Key Stage 3 Reference: Page 33 — *Lonsdale* Science Revision Guides - The Essential Series

COMPOUNDS

Materials and their Properties — 6

1. a) For the following substances indicate whether they are an element or a compound.

SUBSTANCE	ELEMENT OR COMPOUND	SUBSTANCE	ELEMENT OR COMPOUND
Carbon Dioxide, CO_2		Oxygen, O_2	
Sodium Chloride, $NaCl$		Water, H_2O	
Nitrogen, N_2		Copper, Cu	
Argon, Ar		Copper Sulphate, $CuSO_4$	
Neon, Ne		Sulphur Dioxide, SO_2	

b) For the compounds in part a) above, write down the names of the elements each compound contains.

NAME OF COMPOUND	NAMES OF ELEMENTS	NAME OF COMPOUND	NAMES OF ELEMENTS

2. Explain, using diagrams only, how atoms join together to form the following compounds. You must only use 4 metal atoms. Complete the word equation for each one.

a) Sodium Chloride, $NaCl$

[Na Na Na Na] Sodium + [Cl Cl Cl Cl] Chlorine → [NaCl NaCl NaCl NaCl] Sodium Chloride

b) Calcium Oxide, CaO

[ ] + [ ] → Calcium Oxide

c) Potassium Chloride, KCl

[ ] + [ ] → Potassium Chloride

d) Sulphur Dioxide, SO_2

[Sulphur] + [ ] → [ ]

e) Copper Chloride, $CuCl_2$

[ ] + [Chlorine] → [ ]

f) Potassium Oxide, K_2O

[ ] + [ ] → [ ]

MIXTURES

Materials and their Properties — 7

1. a) What is the difference between a mixture and a compound?

   b) The composition of a mixture can vary unlike a compound. Explain why.

2. Here are five diagrams.

   A.   B.   C.   D.   E.

   Which diagram represents:

   (i) One element ☐   (ii) One compound ☐   (iii) Mixture of two elements ☐

   (iv) Mixture of one element and one compound ☐   (v) Mixture of two compounds ☐

3. Are the following substances, mixtures or compounds. Explain your answer.

   a) Rock Salt:

   b) Cola:

   c) Ink:

   d) Cup of Coffee:

   e) Lemonade:

   f) Water in a swimming pool:

   g) Cream:

   h) Magnesium Ribbon:

Key Stage 3 Reference: Page 35 — Lonsdale Science Revision Guides - The Essential Series

SOLUTIONS

Materials and their Properties — 8

1. Emma carried out an experiment to investigate how the solubility of sugar changes with the temperature of the water. Sugar was added to the water until no more dissolved.

Amount of sugar dissolved (grams)	Temperature of water (°C)
20	20
40	30
58	40
81	50
104	60
120	70

a) Describe how Emma would carry out a fair test.

b) Draw a graph of Emma's results.

c) What happens to the solubility of the sugar as the temperature of the water increases?

d) Use the graph to find how much sugar would dissolve at ...

   i) 45°C

   ii) 65°C

e) Use your graph to find the temperature at which ...

   i) 30g of sugar would dissolve:

   ii) 108g of sugar would dissolve:

f) In her investigation which substance was ...

   i) The solute: ii) The solvent:

g) Emma decided to get the sugar back after she had completed the investigation at 70°C.

   i) How much sugar would Emma get back?

   ii) Describe how Emma would get the sugar back.

h) Emma decided to repeat the investigation using flour and not sugar. Explain why this would not work.

2. Emma gets chewing gum on her dress. It will not wash off.
   Other chemicals will get the chewing gum off. Can you explain why?

SEPARATING MIXTURES I - Filtration and Evaporation

Materials and their Properties — 9

1. The diagrams show 3 different ways of separating substances.

   In each of the boxes below write down the letter A, B or C to show which of the methods shown below can be used to separate the mixture.

   - [ ] Sand and water
   - [ ] Dissolved sugar and water
   - [ ] Sand and rocks
   - [ ] Chalk and water
   - [ ] Soil and stones
   - [ ] Dissolved coffee and water
   - [ ] Tea leaves and tea

   A   B   C

2. Some smugglers decided to smuggle some diamonds into the country. The diamonds were small and looked just like sugar. They placed them into a bag of sugar. Customs caught the smugglers. Describe a simple method that customs can use to retrieve the diamonds from the sugar.

3. Sugar is obtained from sugar beet. Sugar is soluble and some organic matter in the sugar beet is insoluble. Describe a simple method of obtaining pure sugar crystals from sugar beet in the laboratory.

4. The diagrams show TWO evaporating dishes. Both contain the same volume of salt solution and both were left to evaporate in the same conditions.

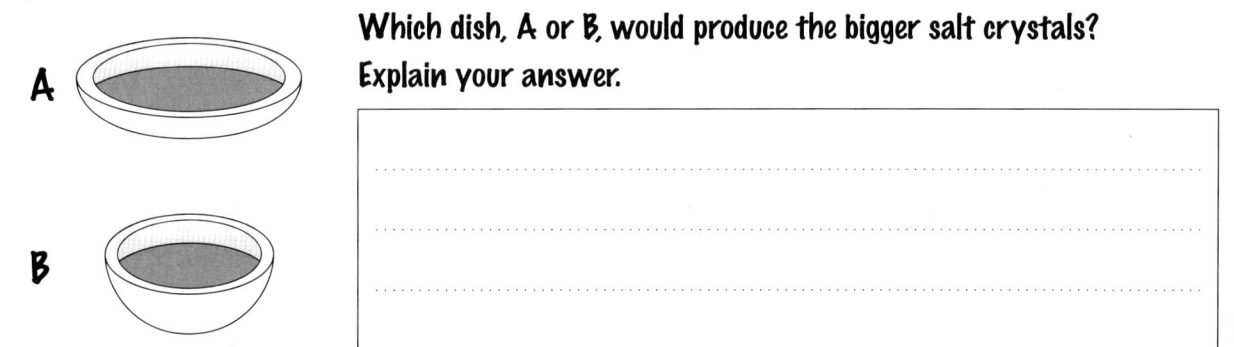

   A

   B

   Which dish, A or B, would produce the bigger salt crystals? Explain your answer.

Key Stage 3 Reference: Page 37

SEPARATING MIXTURES II - Distillation and Chromatography

Materials and their Properties — 10

1. A sample of paint has been taken from the clothes of a hit and run victim. Three suspects with similar coloured cars are under investigation. An experiment has been carried out by the forensics department. The results are shown.

   a) What is the name given to this separation technique?

   Paint taken from victim | SUSPECT A | SUSPECT B | SUSPECT C

   b) i) Explain how the different pigments are carried up the filter paper.

   ii) Why are the pigments carried different distances?

   c) Which suspect is most likely to have committed the crime? Explain your answer.

2. The diagram shows a separation process which can be used to separate a mixture of two solvents. Solvent A boils at 78°C, Solvent B boils at 96°C. The thermometer reads 85°C.

   a) At which point, X or Y, should the cold water enter the condenser?

   b) What is the function of the condenser?

   c) Which solvent, A or B, will be collected in the beaker?

   d) Describe, in detail, how the two solvents would be separated using this apparatus.

   e) Explain how pure water can be obtained from sea water using the apparatus shown.

ROCKS

Materials and their Properties — 11

1. A pupil is given the description of SIX different rocks. For each one name the type of rock and suggest a possible example.

DESCRIPTION	TYPE OF ROCK	EXAMPLE
Grainy, crumbly. Pieces of shell can be seen.		
Dark grey, small crystals, smooth and hard. Split into layers.		
Mainly pale bands of colour, small crystals.		
Small crystals, hard, no layers.		
Crumbly, big grains of sand, orangey colour.		
Crystalline, lots of different coloured interlocking crystals. No layers.		

2. a) What is the name of the process whereby rocks at the surface get broken up?

b) Freeze-thaw is an example of this process as shown below. Explain how the rock gets broken up.

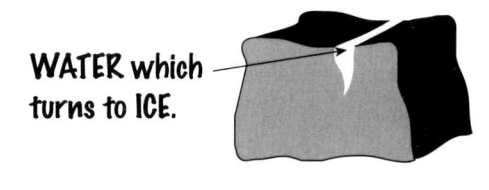

WATER which turns to ICE.

OVER MANY YEARS

c) Describe another process which breaks up rocks at the surface.

Key Stage 3 Reference: Page 39

RECYCLING ROCKS 1

Materials and their Properties — 12

1. The ROCK CYCLE involves three different types of rock which are involved in a continuous, slow UPLIFT and EROSION.

   A. ....
   B. ....
   C. ....
   D. ....
   E. ....
   F. ....
   G. ....

   (SEA, VOLCANO)

   a) Label the diagram using the following:
      IGNEOUS, MAGMA, LAYERS OF SEDIMENT, SEDIMENTARY ROCK, WEATHERED ROCK, UPLIFT TO SURFACE, METAMORPHIC ROCK.

   b) Explain how sedimentary rocks are formed.

   c) Metamorphic rocks are formed from sedimentary rocks. Describe how the change takes place.

   d) Igneous rocks are formed from liquid rock called MAGMA.
      (i) Where does this magma come from? ....
      (ii) What is the change of state when magma crystallises to form igneous rock?

   e) An igneous rock was formed when magma cooled quickly.
      (i) Where would this rock have been formed?

      (ii) Describe one feature of this rock.

      (iii) Suggest a name for this rock. ....

   f) Igneous rock may be changed back into sedimentary rock.
      Four processes involved in this change are listed below. Arrange the processes in order.
      A. Deposition   B. Transport   C. Weathering   D. Burial

      First ☐ → ☐ → ☐ → ☐ Last

RECYCLING ROCKS II

Materials and their Properties — 13

1. Match the following rocks to their rock types.

   LIMESTONE • • IGNEOUS • • SLATE

   SANDSTONE • • METAMORPHIC • • BASALT

   GRANITE • • SEDIMENTARY • • MARBLE

2. The diagram shows a cross-section of a quarry.

   a) i) Why are the crystals of the igneous rock smaller at B than at A?

   ii) How are igneous rocks formed?

   b) i) What type of rock is sandstone?

   ii) Describe how sandstone was formed.

   c) Particles of sand found in sandstone are smooth and round, yet when they were originally broken away from a rock surface they were irregular and jagged. Explain how the shape of the rock has been changed.

   d) (i) There are two different metamorphic rocks formed at points C and D? Explain why.

   (ii) Suggest a name for the metamorphic rocks at point C and at point D.

   C: .... D: ....

Key Stage 3 Reference: Page 41

CHEMICAL REACTIONS

1. When a mixture of iron and sulphur are heated as shown, a compound called iron sulphide is produced.

   POWDERED IRON + POWDERED SULPHUR → ADDED TOGETHER AND MIXED → THEN HEATED → ON COOLING

   a) What type of change is this? Explain your answer.

   b) Write down a word equation for this reaction.

   c) If iron and sulphur are called the reactants, what is iron sulphide called?

2. John heated some calcium carbonate (limestone) in a crucible to form calcium oxide and calcium dioxide. He recorded the mass of the crucible and contents, before and after heating.

   a) Write down a word equation for the reaction that has occured.

   BEFORE Mass = 61.36g
   AFTER   Mass = 60.16g

   b) Why is the mass of the crucible after heating less than its mass before heating?

   c) What does this 'lost' mass represent?

3. Silver oxide is heated in a test tube. A word equation for the reaction is as follows.

   SILVER OXIDE ⟶ SILVER + OXYGEN

   a) i) Name the reactant(s)

   ii) Name the product(s)

   b) If the total mass of silver and oxygen was 10g, how much silver oxide was there to begin with? Explain your answer.

4. Which of the following are chemical changes? Put a tick next to them.

i) A puddle 'drying up' in the sun:	v) Melting butter:
ii) Ice cream melting:	vi) Making a cup of coffee:
iii) Baking a cake:	vii) Frying an egg:
iv) Water condensing on a window:	viii) A car rusting:

BURNING FOSSIL FUELS

Materials and their Properties — 15

1. Butane is a gas which is used as a fuel for camping stoves.

   a) What is a fuel?

   b) Burning butane and all other fuels releases carbon dioxide.
   The table below shows the concentration of carbon dioxide in the atmosphere since 1960.

CONCENTRATION OF CO_2	316	319	323	328	333	338	344	351	358
YEAR	1960	1965	1970	1975	1980	1985	1990	1995	2000

   (i) On the axes below draw a bar graph to show the data.

   (ii) What has happened to the level of carbon dioxide in the atmosphere over the last 40 years?

   (iii) What reasons are there for the changes in atmospheric carbon dioxide levels?

   c) Carbon dioxide is called a 'greenhouse gas.' Explain this statement.

   d) What can we do to reduce the amount of carbon dioxide in the atmosphere?

   e) One of the consequences of global warming is that the polar ice caps are melting. What effect is this likely to have on our country?

2. a) Nitrogen oxides are given off by burning fossil fuels like petrol. There is no nitrogen in petrol. How is nitrogen oxide formed from burning petrol?

   b) How can we reduce the emission of nitrogen oxides into the atmosphere?

Key Stage 3 Reference: Page 43 — Lonsdale Science Revision Guides - The Essential Series

REACTIONS OF METALS

Materials and their Properties — 16

1. The table shows the results of an experiment where FOUR different metals were added to cold water and then to dilute hydrochloric acid.

   a) Place the metals in order of reactivity.

   MOST REACTIVE ....
   ....
   ....
   LEAST REACTIVE ....

METAL	REACTION WITH COLD WATER	REACTION WITH DILUTE HYDROCHLORIC ACID
A	Very vigorous reaction, metal floats and a flame appears.	Cannot be done safely.
B	NO REACTION	NO REACTION
C	NO REACTION	A few bubbles of gas form, slow reaction.
D	Slow reaction, bubbles of gas form.	Reasonable reaction, bubbles of gas form.

   b) Which gas was produced when metal D reacted with dilute hydrochloric acid?

   c) A 'salt' was also produced when metal D reacted with dilute hydrochloric acid. What is a 'salt'?

2. The table below shows the results of an experiment where FIVE different metals were reacted with air.

   a) Which part of the air do the metals react with?

   b) Place the metals in order of reactivity.

   MOST REACTIVE ....
   ....
   ....
   LEAST REACTIVE ....

METAL	REACTION WITH AIR	REACTION WITH WATER	REACTION WITH DILUTE ACID
A	Reacts slowly when heated to form oxide.		
B	NO REACTION		
C	Burns very brightly to form oxide		
D	Burns brightly to form oxide		
E	Reacts very slowly when heated to form oxide		

   c) Complete the table by writing the reactions you would expect these metals to have with water and dilute acid.

3. Why would you not attempt to carry out the reaction between potassium and sulphuric acid?

44   *Lonsdale* Science Revision Guides - The Essential Series   Key Stage 3 Reference: Page 44

THE REACTIVITY SERIES

Materials and their Properties — 17

1. a) Complete the following word equations for the reactions of metals with oxygen. If there is no reaction write 'no reaction'.

   i) Sodium + Oxygen ⟶ ....

   ii) Aluminium + ... ⟶ ....

   iii) Silver + ... ⟶ ....

   iv) Copper + ... ⟶ ....

   b) Use the information above to explain why ...

   (i) After a few months, new shiny copper piping becomes coated with a dark layer. 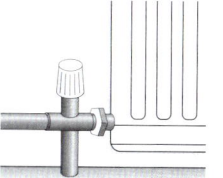

   ....

   (ii) Pure silver rings remain shiny. 

   ....

   (iii) After a few weeks, shiny aluminium becomes coated with a dull layer.

   ....

2. a) Complete the following word equations for the reactions of metals with water. Write 'no reaction' if there is no reaction.

   i) Potassium + water ⟶ ....

   ii) Magnesium + water ⟶ ....

   iii) Iron + steam ⟶ ....

   iv) Copper + Water ⟶ ....

   b) Use the information above to explain why copper hot water pipes are better than aluminium ones?

   ....

3. Complete the following word equations for the reactions of metals with acids. If there is no reaction write 'no reaction.'

   i) Hydrochloric Acid + Calcium ⟶ ....

   ii) Sulphuric Acid + Magnesium ⟶ ....

   iii) Sulphuric Acid + Zinc ⟶ ....

   iv) Hydrochloric Acid + Silver ⟶ ....

   v) Hydrochloric Acid + Iron ⟶ ....

   vi) Sulphuric Acid + Calcium ⟶ ....

Key Stage 3 Reference: Page 45 — *Lonsdale* Science Revision Guides - The Essential Series

DISPLACEMENT REACTIONS

Materials and their Properties — 18

1. When a galvanised zinc nail is placed in a beaker of copper sulphate as shown, the zinc nail becomes coated with a brown solid.

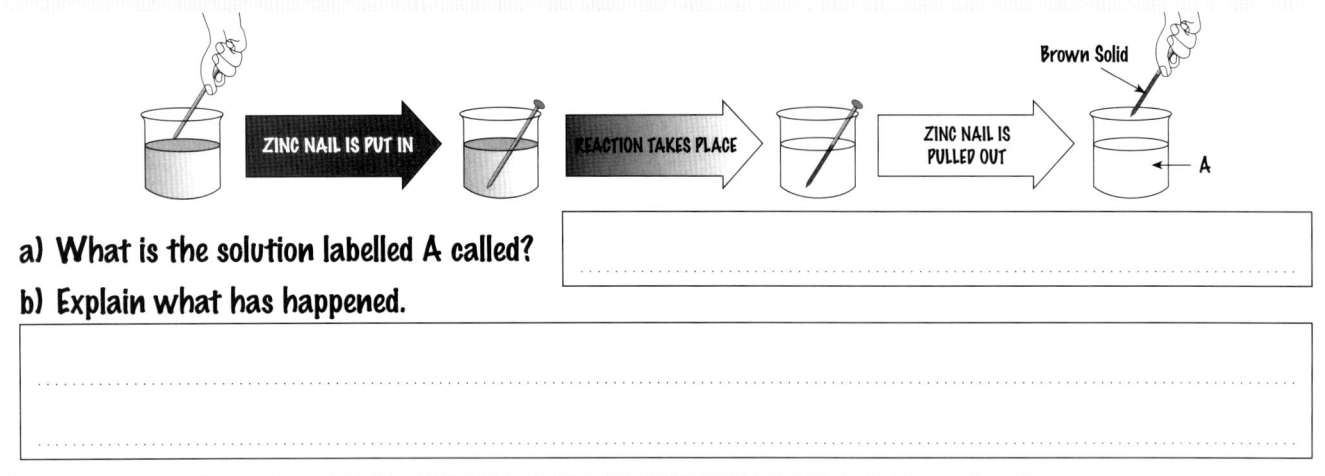

   a) What is the solution labelled A called?

   b) Explain what has happened.

2. Ruth tested SIX metals to see which would react with solutions of different metal nitrates.

   a) Complete the table where ✓ = reaction occurred and ✗ = no reaction occurred.

METAL	Iron Nitrate	Tin Nitrate	Copper Nitrate	Magnesium Nitrate	Lead Nitrate	Zinc Nitrate
Iron						
Tin						
Copper						
Magnesium						
Lead						
Zinc						

   b) Place the SIX metals in order of their reactivity.

   1. ...... 2. ...... 3. ...... 4. ...... 5. ...... 6. ......
   MOST REACTIVE   LEAST REACTIVE

3. For the following combinations of metals and metal compounds complete the word equation for the reaction taking place. If no reaction takes place write 'no reaction.'

   i) SILVER + ZINC NITRATE ⟶ ....

   ii) MAGNESIUM + ... ⟶ .... SULPHATE + IRON

   iii) IRON + COPPER CHLORIDE ⟶ ....

   iv) ZINC + MAGNESIUM SULPHATE ⟶ ....

   v) .... + SILVER NITRATE ⟶ COPPER NITRATE + ...

   vi) ALUMINIUM + ... NITRATE ⟶ .... + LEAD

   vii) MAGNESIUM + ALUMINIUM CHLORIDE ⟶ ....

   viii) CALCIUM + ... ⟶ .... CHLORIDE + SODIUM

   ix) .... + ZINC CHLORIDE ⟶ .... + ZINC

ACIDS AND ALKALIS

Materials and their Properties — 19

1. The table below shows the pH scale. It also shows the range of colours that universal indicator paper would change to in different strengths of acid and alkali.

COLOUR	RED	ORANGE	YELLOW		DARK GREEN	BLUE	DARK BLUE	PURPLE
pH VALUE		5	6	7	8	9	10 - 11	

a) Complete the table.

b) For the following substances state whether they are acidic, neutral or alkaline and write down their pH.

Substance	Colour of universal indicator paper	Acidic, neutral or alkaline	pH
Lemon juice	Orange		
Water	Green		
Oven cleaner	Purple		
Cola	Red		
Salt solution	Green		
Baking powder	Blue		

2. Andy decides to test the pH of soil samples from different parts of his garden. The results are shown in the table. Use the pH scale shown above to answer the questions.

a) Complete the table by writing in the pH of each sample.

b) The vegetable patch needs to be neutral. What would Andy need to add to it? Explain your answer.

PART OF GARDEN	RESULT	pH
North East Plot	GREEN	
South Garden	DARK GREEN	
Vegetable Patch	ORANGE	
North West Herb Plot	GREEN	
Southern Sun Area	RED/ORANGE	
Around the Pond	YELLOW	
Rock Garden	BLUE	

c) (i) Which part of the garden is the most acidic?

(ii) Which part of the garden is the most alkaline?

(iii) Is the south garden acidic, alkaline or neutral? Explain your answer.

d) Heathers grow well in acidic soil, where could Andy place them?

Key Stage 3 Reference: Page 47 — Lonsdale Science Revision Guides - The Essential Series — 47

REACTIONS OF ACIDS

Materials and their Properties — 20

1. Complete the word equations for the following reactions.

    (i) Hydrochloric acid + Magnesium ⟶ .... + ...

    (ii) Sulphuric acid + Zinc ⟶ .... + ...

    (iii) Sulphuric acid + Silver ⟶ .... + ...

    (iv) Hyrdochloric acid + Calcium ⟶ .... + ...

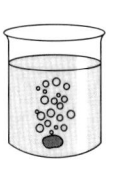

2. Complete the word equations for the following reactions of acids with bases.

    i) .... + Magnesium Oxide ⟶ Magnesium Sulphate + ...

    ii) Hydrochloric Acid + Calcium Oxide ⟶ .... + ...

    iii) Nitric Acid + Zinc Oxide ⟶ .... + ...

    iv) Sulphuric Acid + ... ⟶ Calcium Sulphate + ...

    v) .... + Magnesium Oxide ⟶ Magnesium Nitrate + ...

    vi) .... + Zinc Oxide ⟶ Zinc Sulphate + ...

    vii) Hydrochloric Acid + Copper Oxide ⟶ .... + ...

    viii) Nitric Acid + Sodium Hydroxide ⟶ .... + ...

    ix) .... + Potassium Hydroxide ⟶ Potassium Chloride + ...

    x) .... + ... ⟶ Zinc Chloride + ...

3. Sodium hydrogen carbonate and citric acid are both found in sherbert.

    a) Why does sherbert have a sour taste?

    b) When you put sherbert in your mouth it fizzes.
        i) Explain why the sherbert fizzes.

        ii) What is the name of the gas given off when it fizzes?

4. For the following reactions of acids with carbonates, complete the word equations.

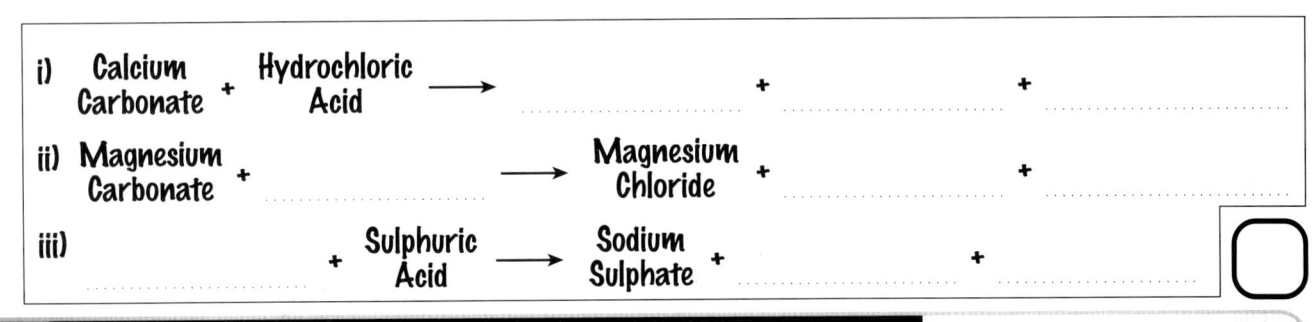

    i) Calcium Carbonate + Hydrochloric Acid ⟶ .... + ... + ...

    ii) Magnesium Carbonate + ... ⟶ Magnesium Chloride + ... + ...

    iii) .... + Sulphuric Acid ⟶ Sodium Sulphate + ... + ...

NEUTRALISATION

Materials and their Properties — 21

1. A piece of apparatus called a burette was used to add alkali to acid in a beaker. Universal indicator solution was added to the acid at the start of the experiment and then the pH was estimated by gauging the colour of the liquid. The liquid was constantly stirred.

VOLUME OF ALKALI ADDED (cm³)	0	16	31	50	65	83	100
pH OF SOLUTION	1	2	3	4	5	6	7

a) Plot these results on the graph paper below.

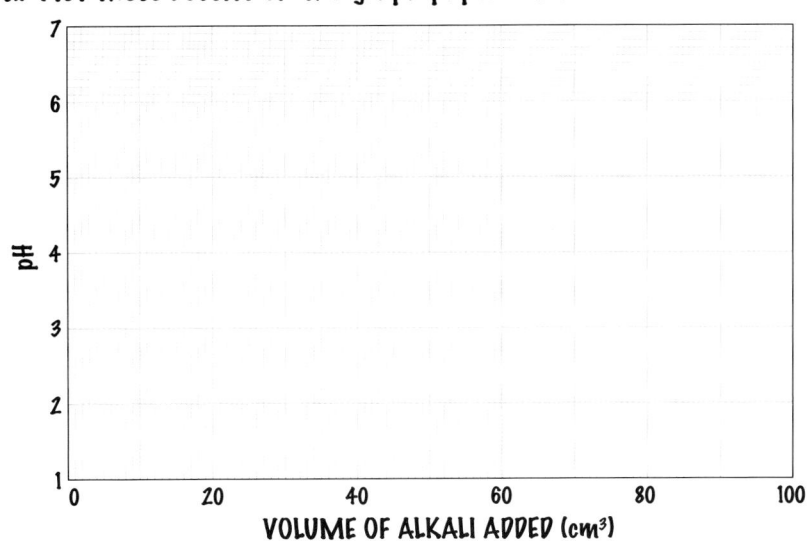

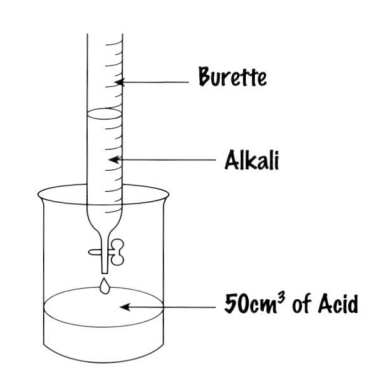

b) What colour was the solution at the start?

c) What was the final colour of the solution?

d) Which was the 'stronger', the acid or the alkali? Explain your answer.

e) What would happen to the pH of the solution if more alkali was added to the solution?

2. The following are all neutralisation reactions. Complete the following word equations.

i) Hydrochloric Acid + Calcium Hydroxide ⟶ ....

ii) Sulphuric Acid + ... ⟶ Sodium Sulphate + ...

iii) .... + Potassium Hydroxide ⟶ Potassium Chloride + ...

iv) .... + ... ⟶ Sodium Chloride + Water

v) Nitric Acid + ... ⟶ Potassium .... + ...

vi) Sulphuric Acid + Calcium Hydroxide ⟶ ....

vii) .... + ... ⟶ Potassium Sulphate + ...

viii) .... + ... ⟶ Sodium Nitrate + ...

ix) Nitric Acid + Calcium Hydroxide ⟶ ....

x) .... + ... ⟶ Magnesium Nitrate + ...

Key Stage 3 Reference: Page 49

ACID RAIN

Materials and their Properties — 22

1. Sulphur dioxide, smoke, carbon monoxide and nitrogen oxides are all air pollutants.

    a) Which TWO of the above pollutants cause acid rain?

    i) .... ii) ....

    b) Name one other gas produced during combustion, which forms an acid when dissolved in water.

    c) How is acid rain produced in the atmosphere?

    d) What is the typical pH of ...

    i) Normal tap water .... ii) Acid rain ....

    e) Statues made of limestone can weather away more quickly than statues made of sandstone, because of the effect of rain water. Explain how rain water makes limestone statues wear away more quickly.

    f) Why has the amount of acid rain increased in the last 200 years or so?

2. The Swedish and other Scandinavian governments frequently complain to the UK government about the effects of acid rain in their country.

    a) What are the environmental effects of acid rain?

    b) Why is it necessary for the Swedish government to protest to the UK government about acid rain?

3. a) Explain how acid rain causes damage to structures such as iron bridges?

    b) Suggest two ways in which acid rain can be reduced.

    i) ....
    ii) ....

Lonsdale Science Revision Guides - The Essential Series — Key Stage 3 Reference: Page 50

BASICS OF ELECTRICITY

Physical Processes — 1

1. For the electrical circuit shown:

   a) What are the names of components A, B, E and F? C and D have been done for you.

   A: ....   B: ....
   C: MOTOR   D: BUZZER
   E: ....   F: ....

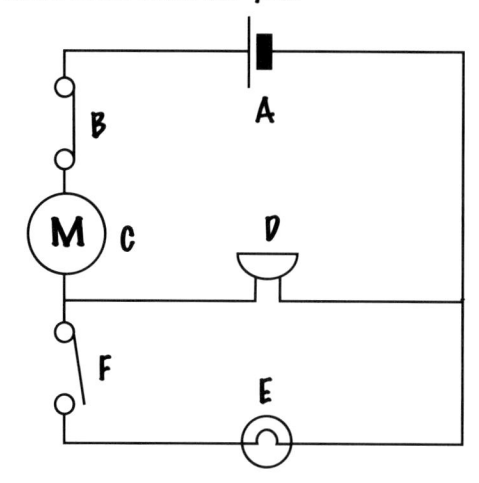

   b) Is this a series or a parallel circuit?

   ....

   c) Which components, if any, are on?

   ....

2. For the following circuits state:

   i) Whether it is a series or a parallel circuit.   ii) Which component or components, if any, are on.

   a) 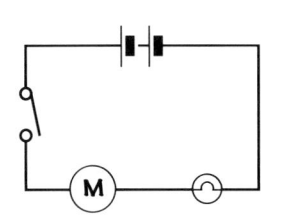
   i) ....
   ii) ....

   b) 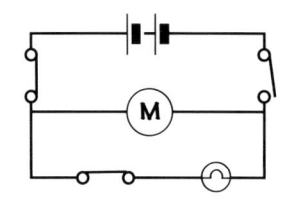
   i) ....
   ii) ....

   c) 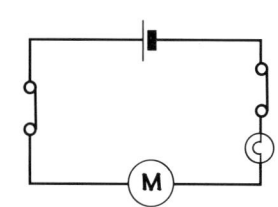
   i) ....
   ii) ....

   d) 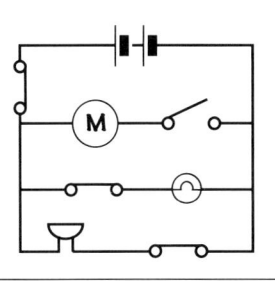
   i) ....
   ii) ....

   e) 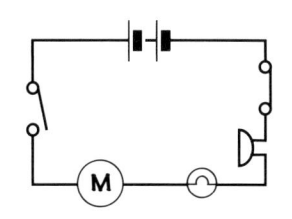
   i) ....
   ii) ....

   f) 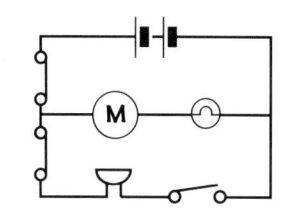
   i) ....
   ii) ....

3. The diagram shows a simple circuit that can be used to control two lamps. Complete the table below.

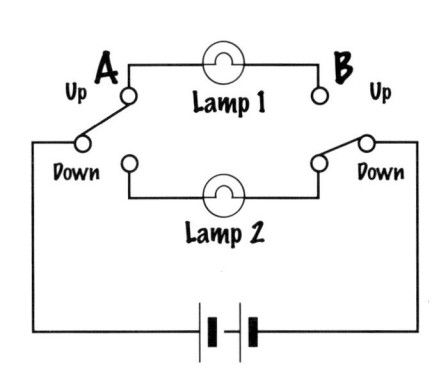

SWITCH A	SWITCH B	LAMP 1 ON or OFF	LAMP 2 ON or OFF
Up	Down		
Up	Up		
Down	Up		
Down	Down		

Key Stage 3 Reference: Page 51 — Lonsdale Science Revision Guides - The Essential Series

ELECTRIC CURRENT

Physical Processes

1. The ammeter reading for a circuit containing one cell and one bulb is 1Amp. For each of the other circuits write down the ammeter reading.

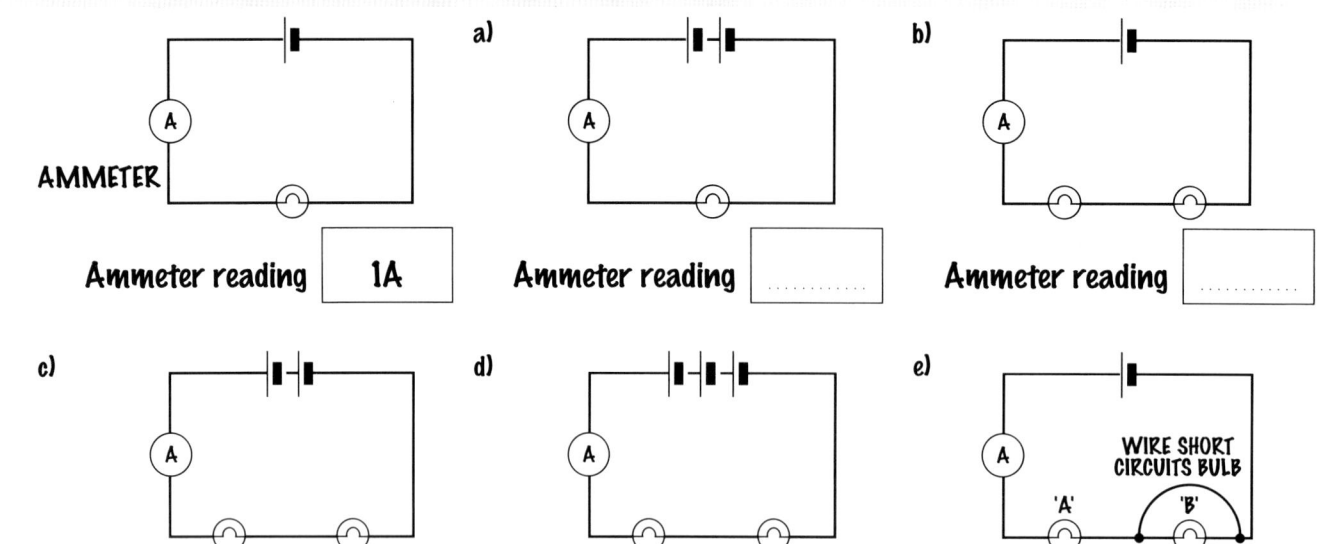

   Ammeter reading: **1A**

   a) Ammeter reading: ....
   b) Ammeter reading: ....
   c) Ammeter reading: ....
   d) Ammeter reading: ....
   e) Ammeter reading (WIRE SHORT CIRCUITS BULB): ....

2. For the electrical circuits below, write down the missing values of current. All the components are identical.

   a) 2A, A_2; A_1
   $A_1 = $ .... $A_2 = $ ....

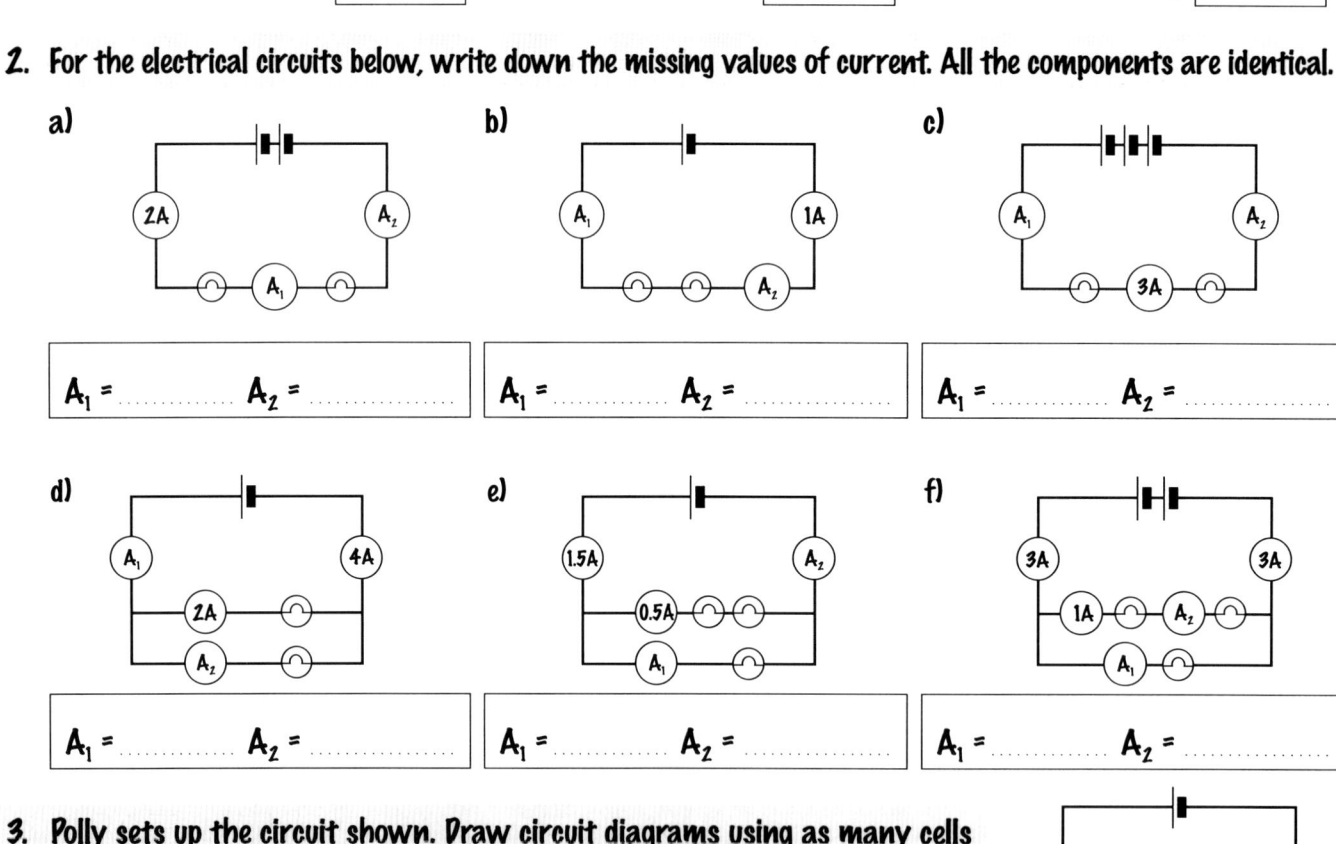

   b) A_1, 1A; A_2
   $A_1 = $ .... $A_2 = $ ....

   c) A_1, A_2; 3A
   $A_1 = $ .... $A_2 = $ ....

   d) A_1, 4A; 2A, A_2
   $A_1 = $ .... $A_2 = $ ....

   e) 1.5A, A_2; 0.5A, A_1
   $A_1 = $ .... $A_2 = $ ....

   f) 3A, 3A; 1A, A_2; A_1
   $A_1 = $ .... $A_2 = $ ....

3. Polly sets up the circuit shown. Draw circuit diagrams using as many cells and bulbs as you want so that the ammeter readings are as follows. (Each cell and bulb are identical)

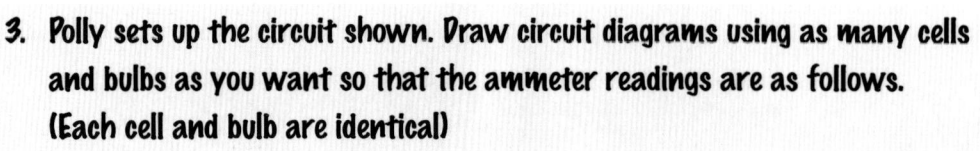

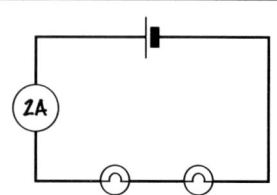

   [Circuit: 2A]

   a) 1A
   b) 3A
   c) 4A
   d) 6A

52   Lonsdale Science Revision Guides - The Essential Series   Key Stage 3 Reference: Page 52

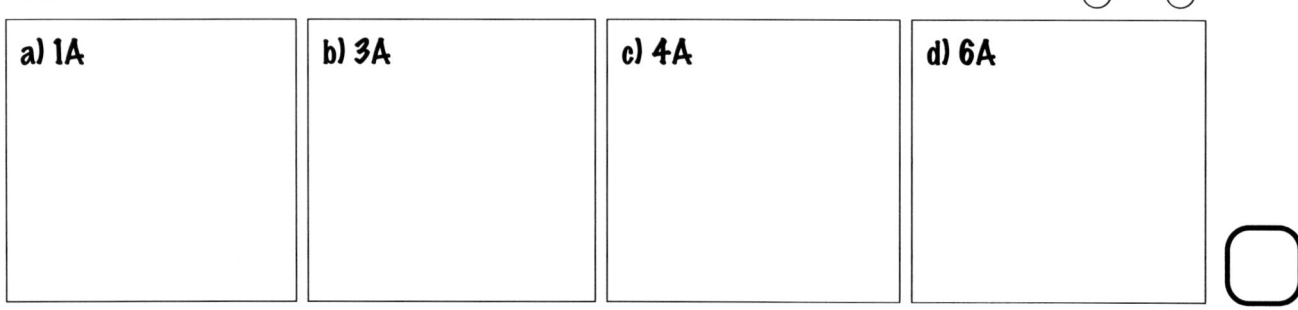

TRANSFERRING ELECTRICAL ENERGY

Physical Processes 3

1. For the following electrical devices describe the energy transfer they are designed to achieve.

   a) Car Headlight: ....
   b) Washing Machine: ....
   c) Electric Kettle: ....
   d) Electric Drill: ....
   e) Electric Cooker: ....
   f) Computer: ....
   g) Computer Printer: ....

2. Which of the following circuits is transferring the most energy every second?

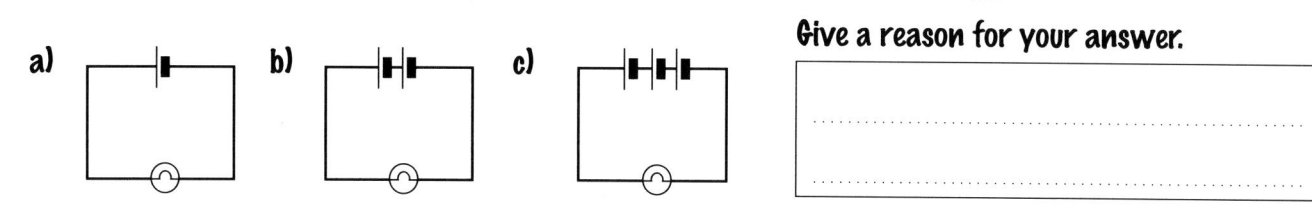

   a)   b)   c)

   Give a reason for your answer.
   ....

3. For each of the circuits shown below draw the position that a voltmeter would need to be placed in to measure the voltage across each lamp.

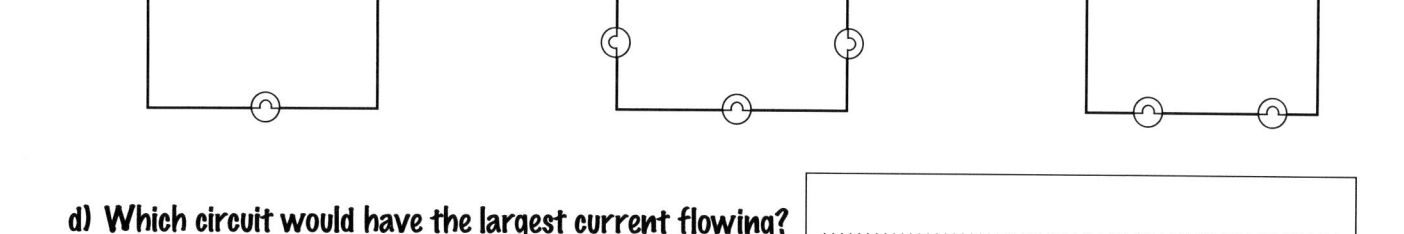

   a)   b)   c)

   d) Which circuit would have the largest current flowing? ....

4. For each of the following circuits write down the values of the missing voltages. All bulbs are identical.

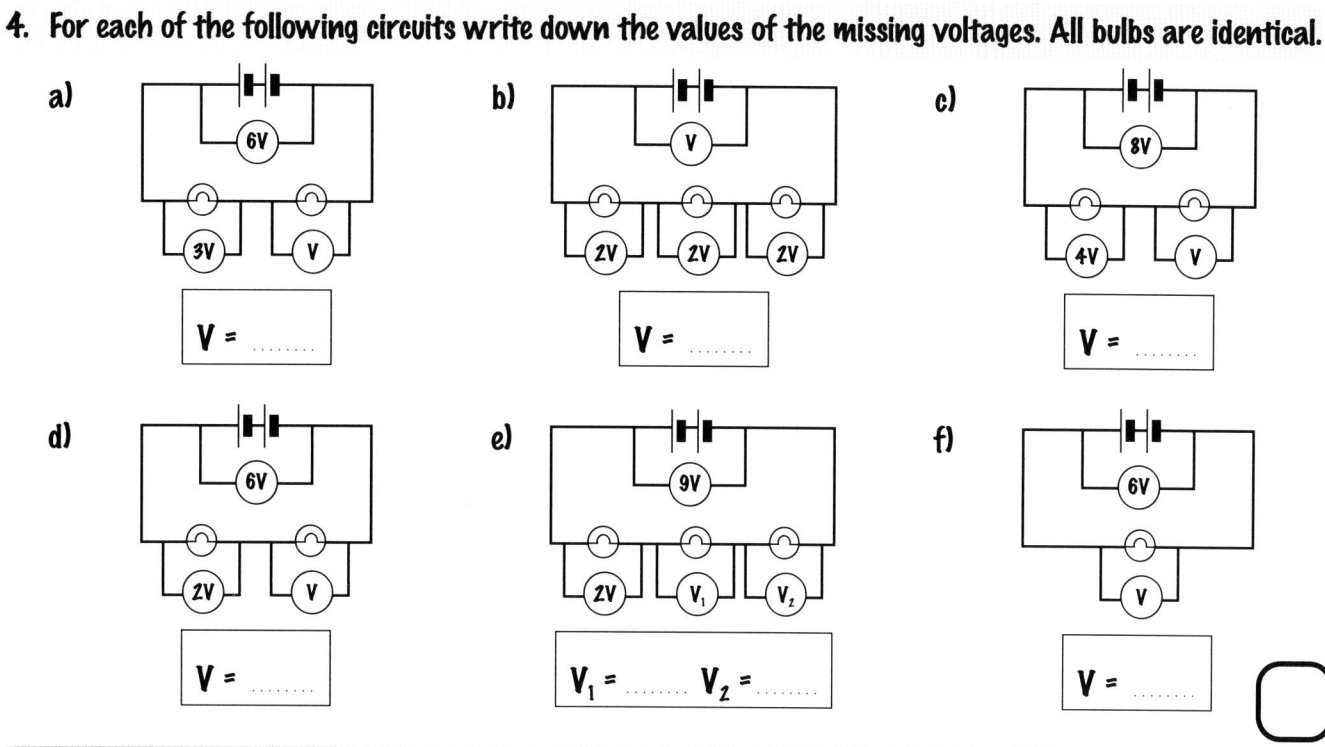

   a) 6V; 3V, V   V = ...
   b) V; 2V, 2V, 2V   V = ...
   c) 8V; 4V, V   V = ...
   d) 6V; 2V, V   V = ...
   e) 9V; 2V, V₁, V₂   V₁ = ... V₂ = ...
   f) 6V; V   V = ...

Key Stage 3 Reference: Page 53

MAGNETIC FIELDS

Physical Processes — 4

1. a) Draw the magnetic field around the single bar magnet shown below.

   b) Write down what we mean by 'magnetic field.'

   c) Write down the names of FOUR magnetic materials.

   (i) .... (ii) ....

   (iii) .... (iv) ....

   d) Name a suitable material for the bar magnet shown.

2. For the magnets below state whether they will attract or repel. Write your answer in the box.

   a)

   b)

   c)

   d)

3. Mary has a bar magnet whose poles are unknown. She also has a magnet whose poles are known. Explain using diagrams how she can determine the unknown poles.

4. The drawing shows a magnetic material being attracted to a bar magnet. When this happens magnetic material itself behaves like a bar magnet.

   a) i) What is the magnetic pole at point A?

   ii) Explain your answer.

   c) A plastic material is now brought close to end A. Explain what happens to the plastic.

54 — Lonsdale Science Revision Guides - The Essential Series — Key Stage 3 Reference: Page 54

ELECTROMAGNETS

Physical Processes — 5

1. Mathew is investigating how the strength of an electromagnet depends on the electric current passing through the coil of the electromagnet.

   a) Name TWO other factors that will affect the strength of the electromagnet.

   i) .... ii) ....

   b) He is going to measure the number of pins the electromagnet can pick up at different current values. How will he carry out a fair test?

   c) The results of the experiment are shown in the table.

CURRENT (A)	1	2	3	4	5	6
No. OF PINS	5	11	15	19	24	30

   i) Plot a graph of his results.
   ii) What conclusion can Mathew come to?

2. The diagram shows a relay.

   a) What is a relay?

   b) Name a suitable material for A.

   c) Explain what happens when the switch is closed.

   d) Suggest ONE possible use for a relay.

Key Stage 3 Reference: Page 55

BALANCED AND UNBALANCED FORCES

Physical Processes — 6

1. What effect do balanced forces have on:

   a) A stationary object?

   b) A moving object?

   c) Unbalanced forces acting on an object cause it to accelerate or decelerate. The greater the difference in the unbalanced forces acting, the greater the acceleration (or deceleration). Diagrams A to F show the forces acting on six moving cars.

   A: 2000N ← car → 3000N
   B: 2500N ← car → 5000N
   C: 2000N ← car → 2500N
   D: 2000N ← car → 2000N
   E: 1000N ← car → 3500N
   F: 2500N ← car → 6000N

   i) Which car has the greatest acceleration? Explain your answer.

   ii) Which car has the smallest acceleration? Explain your answer.

   iii) Which two cars have the same acceleration? Explain your answer.

   iv) Which car is travelling at constant speed? Explain your answer.

2. The diagram shows a basketball player just after throwing a ball to score a basket.

   a) Sketch the path that the ball has followed.
   b) Explain, in terms of balanced and unbalanced forces, the path the ball would have taken.

FORCES IN ACTION

Physical Processes — 7

1. a) The diagram shows a moving car. Name forces A and B.

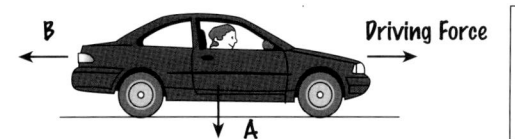

   A: ....

   B: ....

   b) Explain why force B acts on the car.

   ....

   c) The driving force is a result of the tyres NOT being able to spin on the road. What is the name given to the force that stops the tyres spinning?

   ....

   d) The diagram shows a truck moving at the same speed as the car above. Will the car and the truck have ...

   i) The same force A acting? Explain your answer.

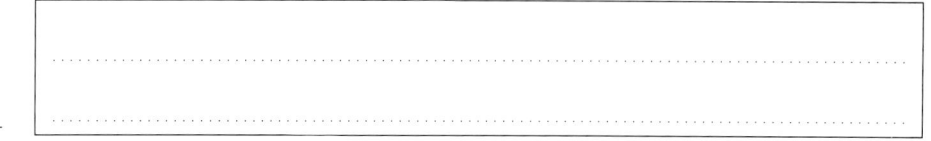

   ....

   ii) The same force B acting? Explain your answer.

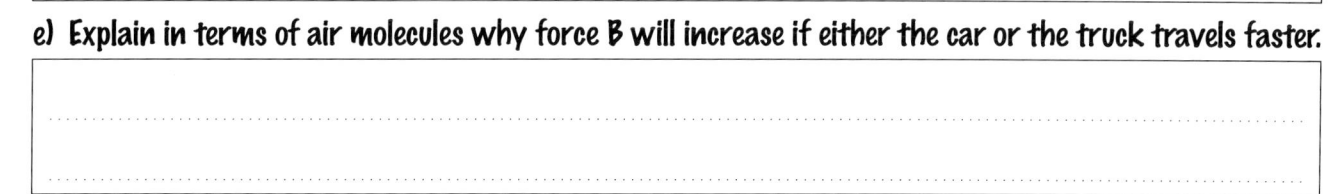

   ....

   e) Explain in terms of air molecules why force B will increase if either the car or the truck travels faster.

   ....

2. The diagram shows two lorries A and B. B has an extra design feature labelled X.

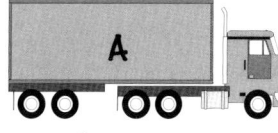

   a) What is the purpose of X?

   ....

   b) Explain how X helps reduce the cost to the environment?

   ....

3. Skiers often 'wax' their skis.

   a) What force do they reduce by doing this? Explain your answer.

   ....

   b) The diagram shows two skiers moving down a slope. Which skier, A or B, would you expect to be moving fastest? Explain your answer.

   ....

Key Stage 3 Reference: Page 57

SPEED

Physical Processes — 8

1. a) A car travels 180km in 2 hours. Calculate the average speed of the car.

   b) An athlete runs at an average speed of 12km/h for 90 minutes. How far does the athlete run?

   c) A cyclist travels at an average speed of 20km/h. Calculate how long it takes the cyclist to travel 50,000m.

2. A cyclist decides to go on a cycling holiday. The holiday lasts 6 days.

   a) Write in the missing values in the table below.

DAY	DISTANCE TRAVELLED IN Km	TIME TAKEN IN HOURS	AVERAGE SPEED IN Km/h
Day 1	75	5	
Day 2		7	15
Day 3	60		5
Day 4		11	10
Day 5	80		8
Day 6	70	5	

   b) What was the total distance travelled during the holiday?

   c) How long did she spend cycling in total?

   d) Calculate her average cycling speed for the holiday.

3. The graph below shows the average speed of a motorist at 10 minute intervals for a journey lasting 90 minutes.

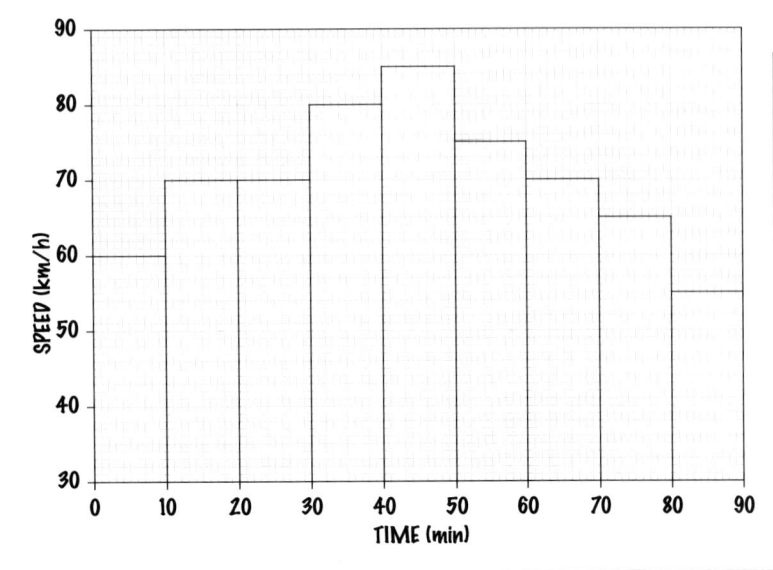

   a) Calculate the average speed of the motorist.

   b) Calculate the distance travelled by the motorist.

PRESSURE

Physical Processes — 9

1. The diagrams show a drawing pin and then a finger being pushed into a notice board with the same force.

   a) Explain why the pin goes into the notice board but the finger does not.

   b) State THREE units of pressure.

   i) .... ii) .... iii) ....

2. Use your knowledge of pressure to explain the following:

   a) Tractors have large tyres: ....

   b) Lorries have as many as eight wheels on their trailers: ....

   c) Football boots have studs: ....

   d) Scissors have sharp blades: ....

3. Calculate the pressure exerted on the ground by each tyre if a car has a weight of 8800N and each tyre has an area of 440cm^2? (Don't forget your formula and your unit).

4. The diagrams below show the outline of Rachel's foot and Derek's foot. Both are drawn on 1cm^2 paper.

   a) Estimate the area of Rachel's foot.

   b) Rachel weighs 500N. Calculate the force she exerts on the ground if she stands on one foot.

   c) Derek stands on one foot and exerts a pressure of 10N/cm^2 on the ground. Calculate his weight.

Key Stage 3 Reference: Page 59

MOMENTS

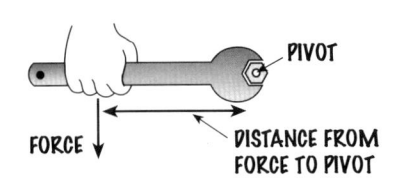

Physical Processes 10

1. a) The girl is trying to unscrew a nut using a spanner. What is the name given to the turning effect of a force.

b) State TWO ways of increasing this turning effect.

i) .... ii) ....

2. The diagrams show two different spanners undoing a nut.

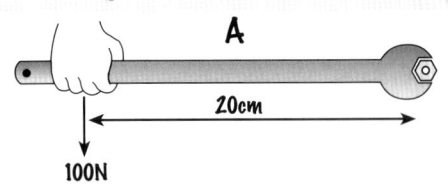

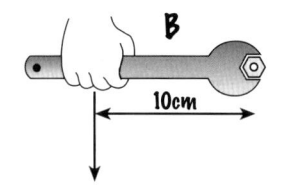

a) Which spanner would make undoing the nut the easiest if the force exerted with both spanners is the same? Explain your answer.

b) For spanner A, calculate the moment of the force exerted.

c) For spanner B, what force is needed to produce the same moment as spanner A?

3. For the following calculate the clockwise moment and the anti-clockwise moment and say whether the seesaw is balanced or unbalanced.

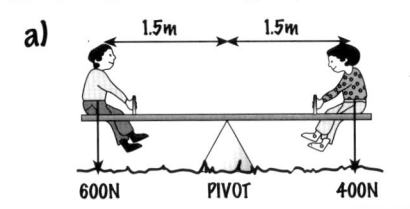

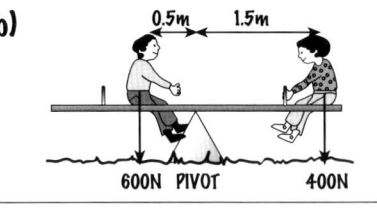

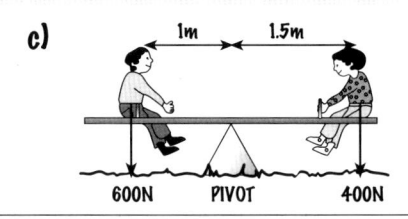

a) 1.5m / 1.5m — 600N PIVOT 400N

b) 0.5m / 1.5m — 600N PIVOT 400N

c) 1m / 1.5m — 600N PIVOT 400N

Clockwise moment =	Clockwise moment =	Clockwise moment =
Anti-clockwise moment =	Anti-clockwise moment =	Anti-clockwise moment =
Balanced/Unbalanced =	Balanced/Unbalanced =	Balanced/Unbalanced =

4. The diagram shows a crane lifting a load. The counter balance has a weight of 10000N.

a) Calculate the moment produced by the counter balance.

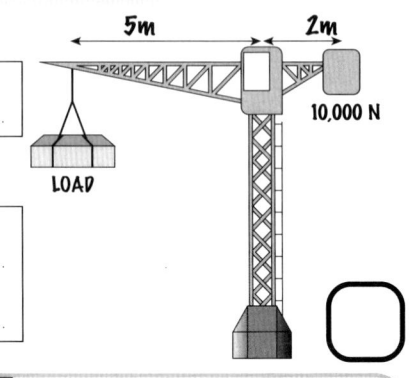

b) If the crane is balanced calculate the weight of the load?

LIGHT

Physical Processes 11

1. Complete the following sentences:

   Light travels in .... at a very high .... A very narrow beam of light is called a .... which is drawn using a .... line with an .... on it to show the .... of travel of the light.

   Objects which give out light are called .... objects. Objects that don't give out light are called .... objects and we can only see them if there is a .... object present.

2. We see objects because light reflects off them.

   a) On the diagram draw the path of a ray of light which enables the girl to see the book.

   b) From the diagram state ONE luminous and ONE non-luminous object.

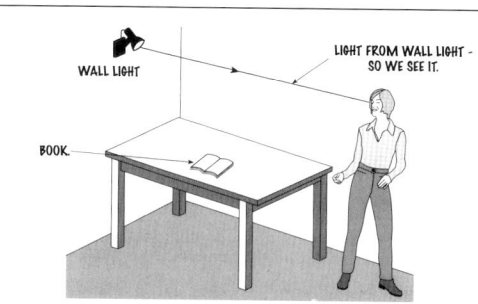

   Luminous: .... Non-luminous: ....

   c) Why is it that you can't see a non-luminous object if there is no luminous source present?

3. A shadow is formed when light travelling from a source is blocked by an object as shown below.

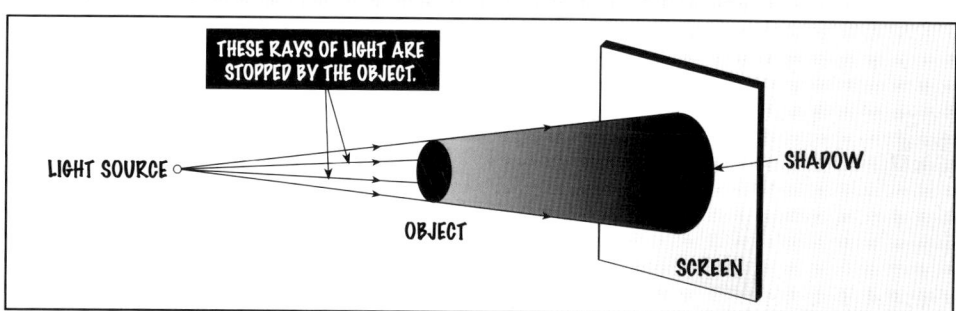

   a) A pupil carried out an investigation to find how the distance from the light source to the object affected the height of the shadow formed. What must the pupil keep the same in order to carry out a fair test.

   b) Plot a graph of the pupil's results.

Distance from light source to object	Height of shadow
10cm	60cm
20cm	30cm
30cm	20cm
40cm	15cm
50cm	12cm

   c) What is the height of the shadow when the distance from the light source to object is 24cm.

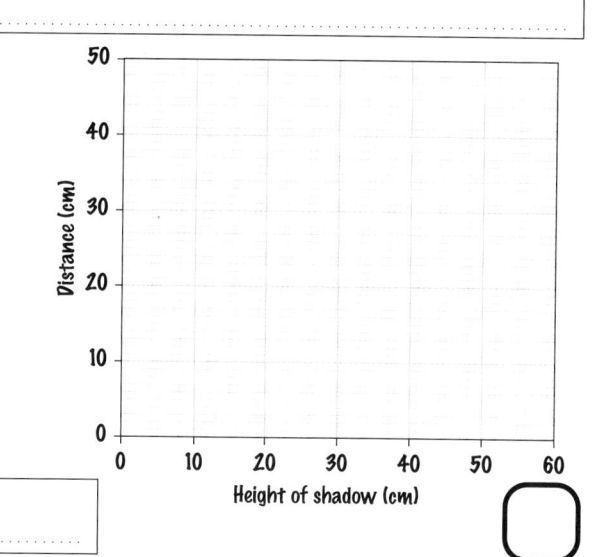

Key Stage 3 Reference: Page 61

REFLECTION OF LIGHT

Physical Processes — 12

1. The diagram shows a ray of light reflected from a plane mirror.

   a) Label the diagram.

   [PLANE MIRROR]

   b) Name the angles 'i' and 'r'.

   i = ... r = ...

2. Using a ruler and a protractor accurately complete this diagram to show how the four mirrors will reflect the ray of light. For each reflection draw a normal and write in the size of the angle of incidence and angle of reflection.

3. For each of the following diagrams write down i) the angle of incidence and ii) the angle of reflection. You are not expected to measure them.

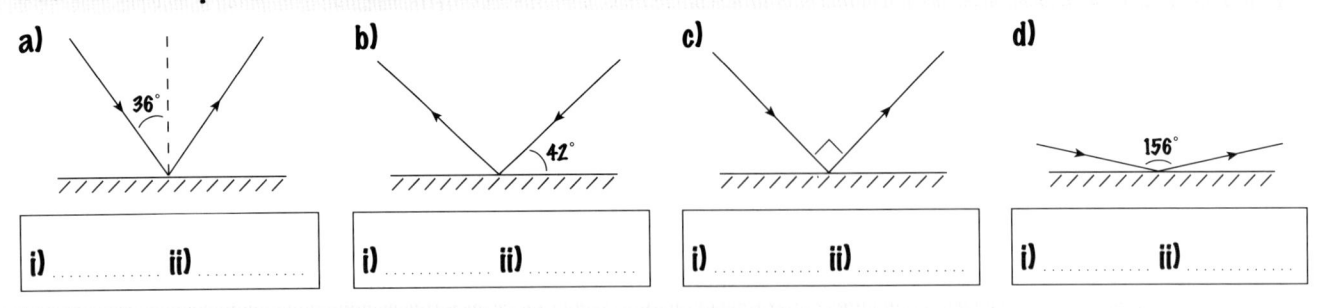

   a) 36°

   b) 42°

   c) (right angle shown)

   d) 156°

   i) .... ii) ....   i) .... ii) ....   i) .... ii) ....   i) .... ii) ....

4. The diagram shows a periscope in use. Draw accurately the path of the ray of light through the periscope.

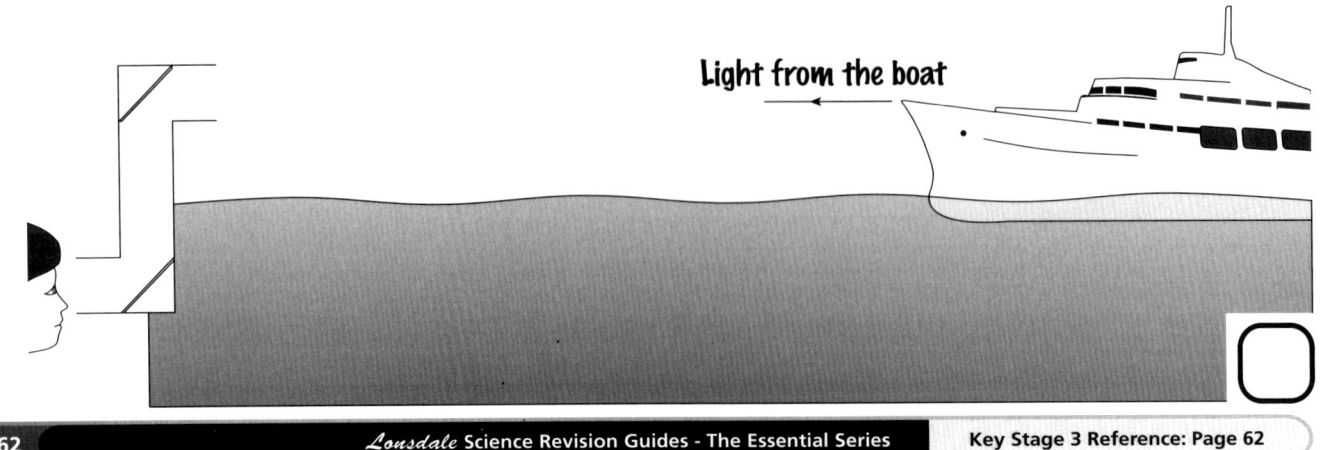

   Light from the boat

REFRACTION AND DISPERSION OF LIGHT

Physical Processes — 13

1. The diagram shows a ray of light passing through a glass block.

   a) When light passes into the glass block it changes direction. What is this effect known as? _____

   b) Why does light change direction when it passes into the glass block?
   _____

   c) Some light does not pass into the glass block. Explain why.
   _____

   d) What happens to the speed of the light as it passes i) into the glass block and ii) out of the glass block?
   i) _____ ii) _____

2. Complete the following diagrams by drawing in all missing rays.

   a)   b)   c)

   d)   e)   f)

3. John places a coin in the bottom of a cup as shown in diagram A.

   a) Draw a ray of light to find out if John can see the coin.
   b) In diagram B John has put lemonade in the cup. John can now see the coin. Draw a ray of light to show how John can see the coin.

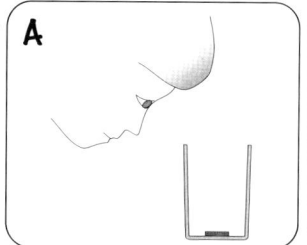

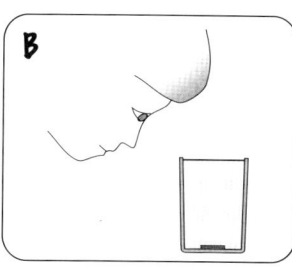

4. The diagram shows what happens when white light passes through a triangular prism.

   a) What is the name given to this 'splitting up' of white light?
   _____

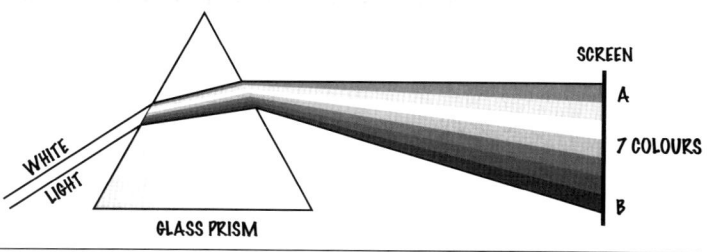

   b) Name the SEVEN colours seen on the screen.

   (i) _____ (ii) _____ (iii) _____ (iv) _____
   (v) _____ (vi) _____ (vii) _____

   c) Which colour is seen at: A = _____ B = _____

Key Stage 3 Reference: Page 63 — *Lonsdale* Science Revision Guides – The Essential Series

COLOUR

Physical Processes — 14

1. Fiona carries out an experiment by shining rays of light at a triangular prism.

   a) Complete the following diagram and then explain what has happened.

   Explanation: ....

   b) Complete the following diagram and then explain what has happened.

   Explanation: ....

2. The red filter in the diagram produces red light. Explain how it does this.

   ....

3. The drawing shows a red filter followed by a blue filter.

   a) What colour will be seen on the screen? ....
   b) Explain your answer.

   ....

4. a) Explain why a green shirt looks green in white light.

   ....

   b) Explain why a green shirt looks black in red light.

   ....

5. Abi is going to the spectrum disco. She's wearing a white top with blue spots and red trousers. She wanders from the blue room to the green room to the red room. What colour will her outfit appear in each room?

   BLUE LIGHT
   Top: ....
   Spots: ....
   Trousers: ....

   GREEN LIGHT
   Top: ....
   Spots: ....
   Trousers: ....

   RED LIGHT
   Top: ....
   Spots: ....
   Trousers: ....

Lonsdale Science Revision Guides - The Essential Series — Key Stage 3 Reference: Page 64

HEARING SOUND

Physical Processes — 15

1. Zed Lep, the rock band, are on their UK tour. Their lead singer is Suzy.

   a) Explain how the sound from the loudspeakers reaches the audience.

   b) Once the sound has reached the audience explain how they hear the sound.

   c) The group play at a loudness of about 140dB. What effect would this have on the ears of the audience?

   d) Explain why many old rock stars are now deaf or partially deaf.

   e) Explain what is meant by 'audible range.'

2. In noisy workplaces ear muffs are worn to reduce the loudness of the sound.

   a) Why do loud sounds damage the ears?

   b) Suggest other ways that noise in the workplace can be reduced.

   c) How is noise reduced inside cars?

3. Complete the wordsearch to find TEN expressions associated with hearing and sound.

D	V	B	N	M	H	V	S	W	R	A	R	E	G	T	F	D	D	E	D	A	H	L	J
Q	W	F	J	L	E	A	R	D	R	U	M	E	Q	R	R	E	S	A	A	D	W	O	K
H	P	I	J	R	R	K	D	D	G	N	V	Y	W	S	H	A	A	R	W	S	J	U	E
V	I	B	R	A	T	I	O	N	H	A	A	R	Q	A	S	F	D	C	A	N	A	D	U
H	W	A	F	H	Z	A	S	V	B	T	J	K	L	O	P	N	G	A	S	O	U	N	D
A	F	R	A	N	G	E	O	F	H	E	A	R	I	N	G	E	T	N	E	I	C	E	S
G	R	A	E	Q	G	K	A	W	X	T	X	N	W	C	H	S	J	A	B	S	X	S	O
A	D	E	C	I	B	E	L	C	L	R	A	D	E	K	Y	S	D	L	G	E	N	S	F

Key Stage 3 Reference: Page 65 Lonsdale Science Revision Guides - The Essential Series 65

VIBRATION AND SOUND

Physical Processes — 16

1. The following statements are all about sound. Tick the ones that are correct.

Sound is a form of energy.	☐	Sound travels faster than light.	☐
Sound can travel through a vacuum.	☐	All sounds are caused by vibrations.	☐
Sound needs a material to travel through.	☐	We hear sounds because our ear drum vibrates.	☐
Humans can hear all sounds.	☐	The loudness of a sound depends on the amplitude.	☐
Loud sounds damage our hearing.	☐	The pitch of a sound depends on amplitude.	☐
Sounds cause the substance they are travelling through to vibrate.	☐	The loudness of a sound depends on the frequency.	☐
Sound is a source of energy.	☐	The pitch of a sound depends on frequency.	☐

2. The diagrams below show the traces of FOUR different sounds.

A, B, C, D

a) Which diagram shows sound with:

i) The biggest amplitude: .... ii) The highest frequency: ....

b) Which diagram shows ...

i) The quietest sound: .... ii) The lowest pitch: ....

iii) The loudest sound: .... iv) The highest pitch: ....

c) Which TWO diagrams show sounds of ...

i) The same pitch but different loudness? ....

ii) Different pitch but the same loudness? ....

3. Massive explosions that emit light and heat occur on the Sun. Explain why we can see the Sun but cannot hear the explosions.

4. At sports day the races are started using a starting pistol. The starting pistol bangs and gives out a puff of smoke. People timing the race see the puff of smoke before they hear the bang. Explain why.

THE SOLAR SYSTEM

Physical Processes — 17

1. Saturn is a planet in the solar system. The Sun is a star.

   a) What is the difference between a star and a planet?

   b) Why does Saturn stay in orbit around the Sun?

   c) Saturn can be seen from Earth. Explain how.

   d) Explain why the position of Saturn in the night sky changes throughout the year.

   e) It takes Saturn 29 years to complete one orbit of the Sun.
      i) Name FIVE planets that take less time than Saturn to orbit the Sun.

      (i) .... (ii) .... (iii) .... (iv) .... (v) ....

      ii) Name THREE planets that take more time than Saturn to orbit the Sun.

      (i) .... (ii) .... (iii) ....

   f) The surface temperature of Saturn is approximately −180°C. Explain why Saturn is colder than the Earth.

2. The table below shows the length of year (compared to the Earth) for the first FIVE planets in the solar system.

PLANET	A	B	C	D	E
LENGTH OF YEAR	0.6	12	0.2	1	1.9

   a) Draw a bar graph to show the information in order of length of year.

   b) Which planet would you expect to be nearest to the Sun? Explain your answer.

   c) Which planet is the furthest from the Sun? Explain your answer.

   d) Name the FIVE planets.

   A) .... B) .... C) ....
   D) .... E) ....

Key Stage 3 Reference: Page 67

THE EARTH

Physical Processes — 18

1. Each day the Sun appears to move across the sky. The diagram shows the position of the Sun at midday on 21st June.

   SUN

   EAST — WEST

   a) On the diagram draw the path the Sun would take from sunrise to sunset.
   b) Draw the path the Sun would take in December, label it D.
   c) Why does the Sun appear to move across the sky?

2. Below are SIX statements which refer to either summer or winter. Link each statement to the correct season.

   - The sun is lower in the sky
   - We are tilted towards the sun
   - Daytime is longer than night-time
   - We are tilted away from the sun
   - The sun is higher in the sky
   - Daytime is shorter than night-time

   SUMMER
   WINTER

3. The diagram shows the position of the Earth relative to the Sun.

   a) Which season is the northern hemisphere experiencing? Explain your answer.

   b) Which season is the southern hemisphere experiencing? Explain your answer.

   c) On the diagram mark with an X a place experiencing night-time.

   d) Explain what would happen to the seasons if the Earth spun around on its imaginary axis as shown in this diagram. Explain your answer.

   e) The diagram shows a 'time exposure' photograph of the night sky.

   • Pole Star

   i) Why is there no movement of the Pole Star?

   ii) Why do the paths of the other stars appear as curved lines?

SATELLITES

Physical Processes — 19

1. a) What is the difference between a natural satellite and an artificial satellite?

   b) Name TWO natural satellites of the Sun.

   i) .... ii) ....

   c) Name the Earth's natural satellite.

2. The diagram shows the possible orbits of TWO artificial satellites.

   a) Which satellite has an orbit which passes over ...

   i) The Equator: .... ii) The poles: ....

   b) How long would a communication satellite take to orbit the Earth?

   c) What is the significance of this time?

   d) The satellite which passes over the poles orbits the Earth several times every day.

   i) What is the advantage of this?

   ii) What is this satellite used for?

3. The Hubble telescope is a satellite used for observing distant galaxies.

   a) Why is it an advantage to have the telescope outside the Earth's atmosphere?

   b) Give ONE disadvantage of having an astronomical telescope in orbit around the Earth.

4. a) The circumference of the orbit of a satellite is 260400km. The satellite is travelling at a speed of 10850km/h. Calculate how long the satellite takes to complete one orbit.

   b) What type of orbit does this satellite have? Explain your answer.

Key Stage 3 Reference: Page 69 — Lonsdale Science Revision Guides - The Essential Series

FORMS OF ENERGY

Physical Processes 20

1. Complete the following crossword. The answers are all forms of energy.

   Across
   1. A climber at the top of a cliff has this type of energy.
   2. Add this type of energy to an object and its temperature rises.
   3. A climber falling down from the top of a cliff has this type of energy!
   4. Hope the climber remembered his vest - it could keep this energy in.
   5. Another name for energy in 3 across

   Down
   6. This kind of energy is used by plants in photosynthesis.
   7. Vibrating objects produce this energy.
   8. This form of energy is produced by plants as a result of photosynthesis.
   9. A generator is a source of this type of energy.
   10. Energy stored by a stretched or compressed object.

2. What form of energy do the following have?

   a) A loaf of bread: ....
   b) A moving car: ....
   c) A diver on a diving board: ....
   d) A bath of hot water: ....
   e) A stretched spring: ....
   f) A cyclist at the top of a hill: ....

3. What energy transfers take place in the following?

   a) A light bulb: ....
   b) A loudspeaker: ....
   c) An electric fire: ....
   d) A battery: ....

ENERGY RESOURCES

Physical Processes — 21

1. a) Coal, oil and gas can be used to generate electricity. Rearrange the following statements in order to show the process

   A. Generators produce electricity.
   B. Boiling water turns to steam.
   C. Burnt fuel produces heat.
   D. Turbines turn the generators.
   E. Steam turns the turbines.
   F. Heat is used to boil water.
   G. Fuel is burned.

   ☐ → ☐ → ☐ → ☐ → ☐ → ☐ → ☐

   b) Wind turbines can be used to generate electricity.

   i) What form of energy does the wind possess?

   ii) Where does this energy come from?

   iii) Wind is a renewable energy resource. Coal is a non-renewable energy resource.
   What is the difference between a renewable and a non-renewable energy resource?

   c) The majority of the world's population use biomass as a fuel.
   i) What is 'biomass'?

   ii) Give ONE advantage of using biomass rather than coal as a fuel.

   iii) Give ONE advantage of using wind rather than coal as an energy resource.

   iv) Give ONE advantage of using coal rather than biomass as a fuel.

2. Why is the Sun considered to be the original source of most energy resources?

3. Explain why a battery is a convenient energy resource.

Key Stage 3 Reference: Page 71 — Lonsdale Science Revision Guides - The Essential Series

TEMPERATURE AND HEAT

Physical Processes — 22

1. Petra and Ezra carried out an experiment. They measured the temperature of a can of cold drink and a warm cup of tea every minute for 10 minutes.

   a) What did they use to measure the temperature?

   b) Draw two graphs to show their results.

Time (min)	TEMPERATURE (°C) Can	TEMPERATURE (°C) Tea
0	4	44
1	8	36
2	11	30
3	13	26
4		
5	16	21
6	17	20
7	18	20
8	19	20
9	20	20
10	20	20

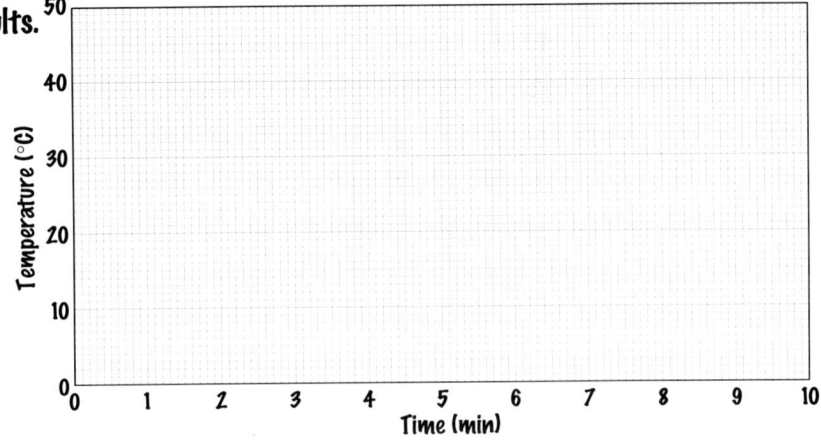

   c) Ezra forgot to write down the results after 4 minutes. Use the graphs to estimate the temperatures at this time.

   Can: .... Tea: ....

   d) Ezra says that the graphs show that liquids cool down quicker than they warm up. Petra disagrees. Is there enough evidence from these results to support Ezra's argument? (Think about fair tests)

   e) Explain why the can warms up and the tea cools down.

2. Ezra decides he wants to carry out an experiment to find out whether a blue Bunsen Burner flame or a yellow one gives out most heat energy.

   a) Design a fair test, including a diagram, that Ezra could use to measure which flame gives out the most heat energy?

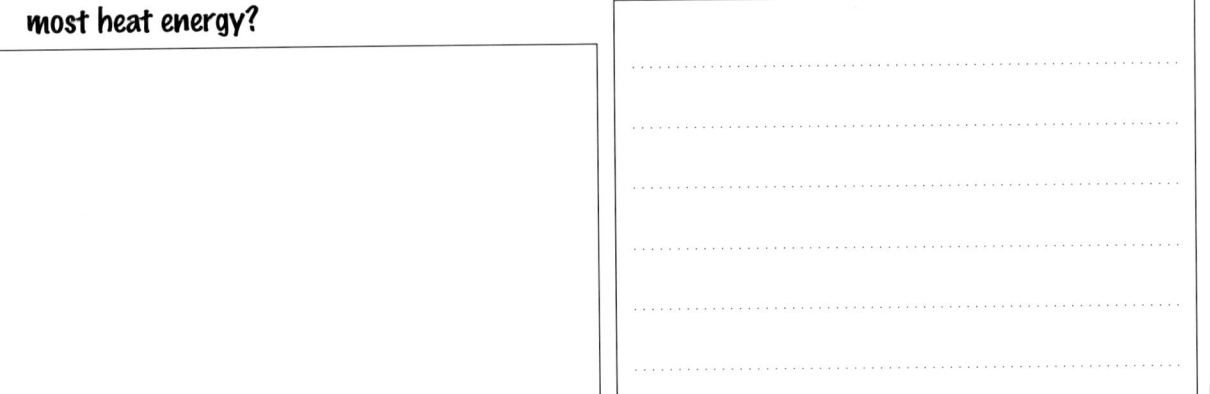

CONDUCTION OF HEAT

Physical Processes — 23

1. Cathy and Jon are carrying out an experiment. Jon is holding a copper rod into a Bunsen flame and Cathy is holding a glass rod.

   a) Which person feels the heat first? Explain your answer.

   b) Explain what is meant by 'conduction of heat.'

2. a) Phil has lost the grips on the handlebars of his bike. Explain why they feel colder than normal.

   b) Explain why birds 'fluff-up' their feathers in cold weather.

   c) Sarah has a feather filled duvet. Why does she always give it a good shake on cold winter nights?

   d) Explain why a 'fleece' is good at keeping you warm.

3. Use your knowledge of conductors and insulators to name a suitable material for the following and give a reason for choosing it.

   a) Car radiators:
   b) Cool boxes:
   c) Lining a fridge:
   d) Container to keep a burger hot:
   e) Bottom of a pan:
   f) Handle of a pan:

4. Ruth and Mel go on a picnic. Ruth uses a cool bag to keep the lemonade cool. Mel uses a cool bag to keep the sausage rolls warm. Explain why a cool bag is suitable for both purposes.

Key Stage 3 Reference: Page 73 — *Lonsdale* Science Revision Guides - The Essential Series

CONVECTION, RADIATION AND EVAPORATION OF HEAT

1. a) The diagram shows a 'convector' heating a room. Use arrows to show the passage of the air around the room.

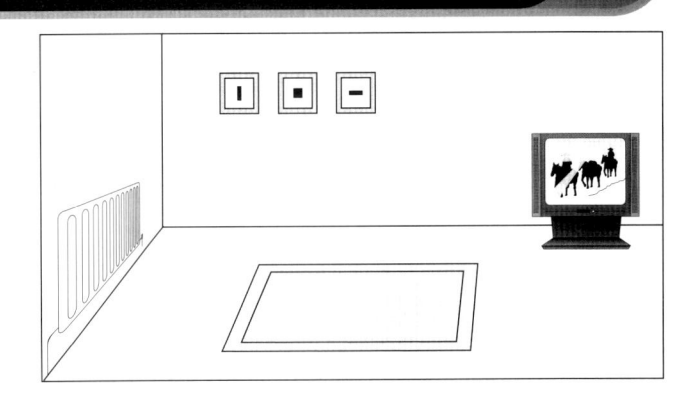

b) Explain how a coal fire helps to ventilate a room.

2. Ahmed and Jim are carrying out an experiment to find out whether a can with a dull black surface or a bright shiny one is the best radiator of heat. They put hot water in each can and then took its temperature every minute for ten minutes.

a) Write down how Ahmed and Jim should carry out a fair test.

b) Their results are shown below.

Temperature (°C)	Bright Shiny Can	80	73	66	59	54	50	45	40	38	37	36
	Dull Black Can	80	71	63	56	50	45	41	38	36	34	32
	Time (min)	0	1	2	3	4	5	6	7	8	9	10

Plot two graphs to show the results. Label each graph.

c) What conclusion can Ahmed and Jim come to?

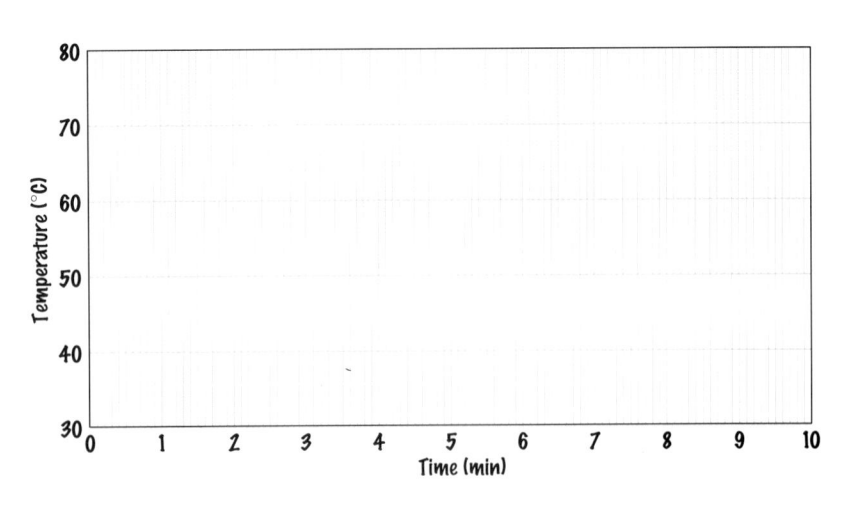

3. Explain why changing rooms in swimming baths are always kept very warm.

TRANSFER AND CONSERVATION OF ENERGY

Physical Processes — 25

1. For the following electrical devices write down the <u>main</u> energy transfer.

   a) A bicycle dynamo transfers .... energy to .... energy.
   b) A battery transfers .... energy to .... energy.
   c) A microphone transfers .... energy to .... energy.
   d) An electric cooker transfers .... energy to .... energy.
   e) A hairdryer transfers .... energy to .... energy.

2. a) What energy conversion is the vacuum cleaner designed to bring about?

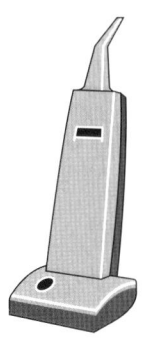

   .... energy ⟶ .... energy

   b) If 1000J of energy are input to the motor and 600J of energy is output as useful energy, how much energy is wasted?

   c) What form does the wasted energy take?

3. a) What form of energy does a motor transfer?

   b) The table below shows the useful and wasted energy produced by a motor every second.

Energy Produced	No. of joules produced every second
Kinetic	40
Heat	35
Sound	25

   i) How much energy does the motor transfer every second?

   ii) Draw a bar graph to show the data in the table.

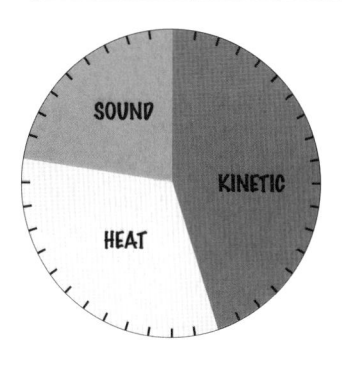

   c) The pie chart shows the useful and wasted energy transfer per second for a different motor. 90 Joules of kinetic energy is produced by the motor every second.

   i) How much heat energy does the motor produce every second?

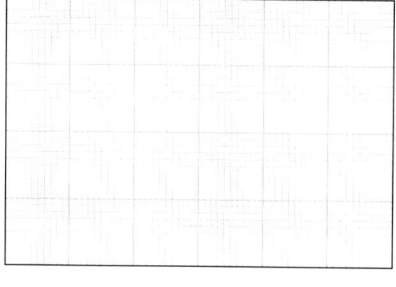

   ii) How much sound energy does the motor produce every second?

   iii) How much energy does the motor transfer every second?

   d) Which motor do you think is the most efficient? Explain your answer.

Key Stage 3 Reference: Page 75 — *Lonsdale* Science Revision Guides – The Essential Series

hello, why dosen't this book have answers in it, I don't know wat I got right + wrong.

NOTES

NOTES

THE COMPLETE KEY STAGE 3 PACKAGE ...

CONTENT ...

- Full colour.
- Differentiated worksheets for each page of content available on our website. 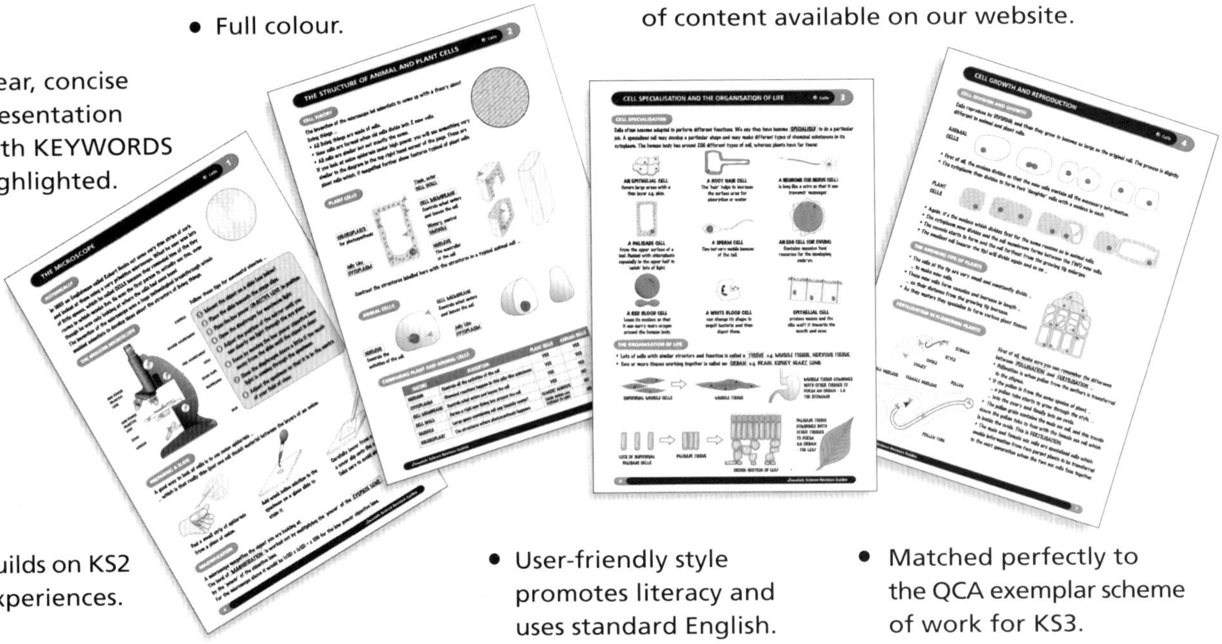
- Clear, concise presentation with KEYWORDS highlighted.
- Builds on KS2 experiences.
- User-friendly style promotes literacy and uses standard English.
- Matched perfectly to the QCA exemplar scheme of work for KS3.

... REINFORCEMENT ...

- Develops vocabulary by matching KEYWORDS exercise in every unit.
- Develops reading and understanding by comprehension exercises on scientific ideas. 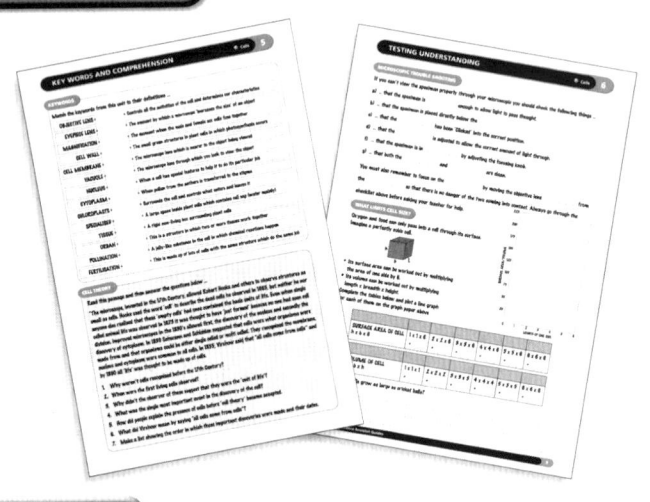
- Exercises covering various aspects of the content can be used for class-work or homework.

... INVESTIGATION.

- An investigation for each unit.
- Planned so as to be achievable within the timescale of the unit. 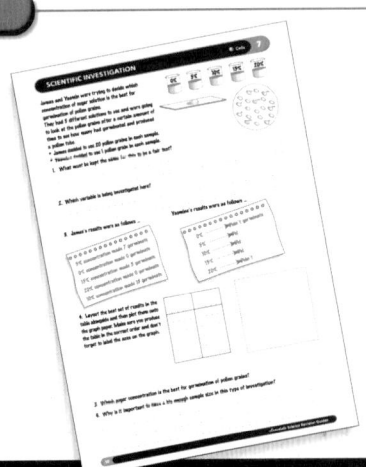
- Develops investigative skills by focusing the pupils on performing 'fair' tests.
- Could be carried out as a true practical investigation or treated hypothetically by providing data.

Plus...

... 300 PAGES OF DIFFERENTIATED WORKSHEET SUPPORT ON OUR WEBSITE. WWW.LONSDALESRG.CO.UK/WEBSUPPORT (FROM SEPT. 15TH)

Lonsdale Science Revision Guides

... IN FULL COLOUR.

ONE COMPLETE COURSE GUIDE ...

- Cells
- Reproduction
- Environment & Feeding Relationships
- Variation & Classification
- Acids & Alkalis
- Simple Chemical Reactions
- Particle Model of Solids, Liquids & Gases
- Solutions
- Energy Resources
- Electrical Circuits
- Forces and Their Effects
- The Solar System and Beyond

... FOR EACH YEAR OF ...

- Food & Digestion
- Respiration
- Microbes & Disease
- Ecological Relationships
- Atoms & Elements
- Compounds & Mixtures
- Rocks & Weathering
- The Rock Cycle
- Heating & Cooling
- Magnets & Electromagnets
- Light
- Sound & Hearing

... THE KEY STAGE 3 COURSE

- Inheritance & Selection
- Fit & Healthy
- Plants & Photosynthesis
- Plants for Food
- Reactions of Metals & Metal Compounds
- Patterns of Reactivity
- Environmental Chemistry
- Using Chemistry
- Energy & Electricity
- Gravity & Space
- Speeding Up
- Pressure & Moments

Published in May 2001. This year's Year 9 pupils should use our Essentials of Science Key Stage 3 Revision Guide and its associated workbook.

Lonsdale Science Revision Guides

Lonsdale SCIENCE REVISION GUIDES

The bestselling Key Stage 3 Revision Guide in FULL COLOUR ...

... now fully revised to match the new Programme of Study.
Only £2.00 for school orders.
£2.50 for private individuals.